D0896681

THE GOLDEN SECTION

Garth E. Runion

DALE SEYMOUR PUBLICATIONS

For permission to reproduce the photographs on the pages indicated, acknowledgment is made to the following:
Illustrations on page 88 and 89 taken from *The Geometry of Art and Life* by Matila Ghyka, 1946, Sheed & Ward Inc., New York; Flower on page 92, courtesy of Arnold Ryan Chalfant; Starfish on page 92, courtesy of Fred Ward – Black Star; Chambered nautilus on page 95, courtesy of Eastman Kodak Company

Order number DS01033
ISBN 0-86651-510-0

DALE
SEYMOUR
PUBLICATIONS
P.O. BOX 10888
PALO ALTO, CA 94303

Preface

As a mathematics teacher, I know that many of us are always looking for interesting topics that will appeal to students of varying abilities. To meet this need, I have written this book—for certainly there is something for everyone in a study of the Golden Section.

The Golden Section, also known as the "Golden Mean" and the "Divine Proportion," is a ratio found in art and nature that has fascinating mathematical properties. This book explores these geometric and algebraic properties in a variety of challenging activities.

Some students will delight in the construction problems or in creating original designs using the pentagon and pentagram. Others will be challenged by working through the proofs or studying the appendices. Students and teachers will find even greater challenges—and corresponding rewards—in the references at the end of the book.

The book's second purpose is to provide basic information on the Golden Section, a topic that has interested both mathematicians and artists for several centuries. It seems unfortunate that such a topic has been all but dropped from the high school mathematics curriculum. I am not suggesting that students be required to study the Golden Section in detail; I do think, however, that they will profit from the many new and intriguing ideas in this book. Furthermore, the book is a rich source of problem solving activities for junior and senior high school students of differing abilities.

How you use this book depends on your resourcefulness and priorities. You need not limit its use to the classroom—students studying independently can use the material effectively.

The book includes eleven sections and four appendices. Appendix D, which contains nine challenging new problems, has been added to the three appendices included in the first edition. The problems in the book are most appropriate for those with some background in high school geometry. One- and two-starred problems are generally more difficult or time-consuming than the other problems.

For maximum benefit, students should work through the sections in sequence, particularly if they are working independently. I would encourage readers of this book to do the problems; those who do not will miss a great deal.

I hope that both student and teacher will find this book interesting and enjoyable. A special word of appreciation is in order to Mrs. Connie Mignonne Barton Evans for her invaluable assistance in preparing this manuscript.

Contents

To Laura, Amy, Mindy,
Garth II, and Amanda

SECTION 1

What Is
The
Golden Section?

For centuries mathematicians, artists, and nonmathematicians
have been fascinated by this "thing" called the *Golden Section,*
the *Golden Mean,* the *Divine Proportion,* the *Divine Section,*
the *Golden Ratio*, and the *Golden Proportion.* These terms are
all synonymous, and refer simply to the division of a segment
into what is called *extreme and mean ratio.* To understand the
concept of the Golden Section, therefore, it is only necessary
to understand what is meant for a segment to be divided into
extreme and mean ratio. The concept of extreme and mean ratio

is dependent upon the following definition of mean proportional.

> **If a, b, and c are positive numbers such that $a^2 = b \cdot c$, then a is the *mean proportional* of b and c.**

For example, what is the mean proportional of 2 and 3? Let x be the mean proportional of 2 and 3. By our definition, $x^2 = 2 \cdot 3$, or $x = \sqrt{6}$.

Another way of indicating that one number is the mean proportional of two others is to use a simple proportion. Referring to the example above, we could write $\dfrac{2}{\sqrt{6}} = \dfrac{\sqrt{6}}{3}$. In general, if a is the mean proportional of b and c, then $\dfrac{b}{a} = \dfrac{a}{c}$.

With this definition in mind, let us consider a specific segment, $\overline{AB}$, shown in Figure 1. (Throughout this book, notation such as $\overline{AB}$ will be read "segment AB." Such notation refers to the set of points that make up the particular segment. The notation AB stands for the real number that is the length of $\overline{AB}$.)

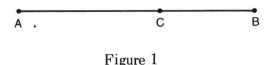

A   C   B

Figure 1

Point C is said to divide $\overline{AB}$ into extreme and mean ratio if and only if one of the lengths AC or CB is the mean proportional of AB and the other length. In Figure 1, therefore, if C divides $\overline{AB}$ into extreme and mean ratio, we know that *either* the proportion $\dfrac{AB}{AC} = \dfrac{AC}{CB}$ (i.e., AB $\cdot$ CB = AC²) *or* the proportion

$\dfrac{AB}{CB} = \dfrac{CB}{AC}$ (i.e., $AB \cdot AC = CB^2$) is true. Although one cannot tell just by looking, point C in Figure 1 has been located so as to divide $\overline{AB}$ into extreme and mean ratio. Thus, Figure 1 is an example of the Golden Section. In the next section, you will learn how to construct a Golden Section.

PROBLEMS ???

1 What is the mean proportional of each of the following pairs of numbers?
   a) 2, 8   **b)** 4, 9   **c)** 5, 9   **d)** $\sqrt{3}, \sqrt{27}$

2 Simplify the following radicals.
   a) $\sqrt{24}$ **b)** $\sqrt{32}$ **c)** $\sqrt{41}$ **d)** $\sqrt{125}$

3 In the last paragraph of this section, it was mentioned that with respect to Figure 1, *either* $\dfrac{AB}{AC} = \dfrac{AC}{CB}$ or $\dfrac{AB}{CB} = \dfrac{CB}{AC}$. Assuming that the lengths of $\overline{AC}$ and $\overline{CB}$ are what they appear to be (that is, AC > CB), which of the two proportions must be true in this case? Explain why the other proportion cannot hold.

4 Suppose that a point F divides $\overline{KL}$ into extreme and mean ratio such that KF < FL. Write the proportion that would indicate this fact. Indicate this fact in another way.

5 What is the division of $\overline{KL}$ into extreme and mean ratio called?

6 Assume that $\dfrac{f}{g} = \dfrac{g}{m}$ and that f, g, $m > 0$.

   a) Find f when $g = 3$ and $m = 7$.
   b) Find g when $f = 5$ *and* $m = 15$.
   c) Find m when $f = 6$ and $g = 4$.
   d) Find g when $f = 2t$ and $m = 8t$. Assume $t > 0$.
   e) Find f when $g = 3x$ and $m = 6z$. Assume x, $z > 0$.

7 In the definition of mean proportional given in this section, what is a if $b = c$? Show that a is always between b and c when $b \neq c$.

8 In geometry you learned that if $\triangle ABC$ is a right triangle, then the length of the altitude drawn to the hypotenuse is the mean proportional of the lengths of the two segments formed on the hypotenuse. This is illustrated by the sketch below.

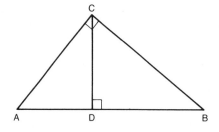

$$\frac{AD}{CD} = \frac{CD}{DB}$$

   a) Find CD if AD $= 2$ and DB $= 8$.
   b) Find CD if AD $= 1$ and AB $= 10$.
   c) Find AD if CD $= 7$ and DB $= 7$.
   d) Find CD if AD $= \sqrt{2}$ and DB $= \sqrt{8}$.

9 Prove the following for a, b, and c positive numbers.

   a) If $a^2 = b \cdot c$, then $\dfrac{b}{a} = \dfrac{a}{c}$.

   b) If $\dfrac{b}{a} = \dfrac{a}{c}$, then $a^2 = b \cdot c$.

SECTION 2

Constructing
The
Golden Section

In the previous section we defined the Golden Section as the division of a line segment into extreme and mean ratio. Now we will see how to locate a point geometrically on a given segment that will divide it into extreme and mean ratio. In the Euclidean tradition, the only instruments we will use are a straightedge and compass. Although there are many different ways to perform this construction, we will consider just one here. However, two alternate approaches are given in Appendices A and B.

To begin the construction, draw $\overline{AB}$. Next, on $\overline{AB}$ construct a perpendicular at B. Now bisect $\overline{AB}$. Call M the midpoint of $\overline{AB}$. On the perpendicular you constructed at B, locate a point N such that BN = AB. Using M as the center and MN as the radius, draw a semicircle intersecting $\overline{AB}$ extended at points R and S. With your compass determine the length of $\overline{BS}$ and locate T on $\overline{AB}$ such that AT = BS. Point T then divides $\overline{AB}$ into extreme and mean ratio.

Your drawing should look like the one in Figure 2.

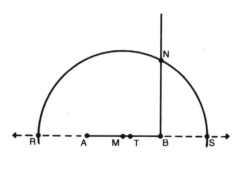

Figure 2

When doing constructions in mathematics it is a good idea, upon completion of the construction, to validate the construction process. This means that by using geometric and algebraic properties we mathematically prove that the construction does what it is supposed to. In the construction in Figure 2, what should we do in order to prove that T really does divide $\overline{AB}$ into extreme and mean ratio? From Section 1, we know that if we can show $AT^2 = AB \cdot TB$, then our construction is correct. If we can show that the simple proportion $\dfrac{AB}{AT} = \dfrac{AT}{TB}$ follows from what we did in our construction, then we know our procedure is correct. The proof in this case turns out to involve two theorems from plane geometry.

Theorem 1: An angle that is inscribed in a semicircle is a right angle.

Theorem 2: The length of the altitude that is drawn to the hypotenuse of a right triangle is the mean proportional of the lengths of the two segments formed on the hypotenuse. (See Section 1, Problem 8.)

Referring again to Figure 2, we notice that since $\overset{\frown}{RNS}$ (read "arc RNS") is a semicircle, if we were to draw $\overline{RN}$ and $\overline{NS}$, triangle RNS would have, by Theorem 1, a right angle at N. This would mean that $\overline{NB}$ is the altitude to the hypotenuse. Hence, by Theorem 2, $\dfrac{RB}{BN} = \dfrac{BN}{BS}$. It follows then that

$\dfrac{RB}{BN} - 1 = \dfrac{BN}{BS} - 1$, or that $\dfrac{RB}{BN} - \dfrac{BN}{BN} = \dfrac{BN}{BS} - \dfrac{BS}{BS}$. From the last

equation, we can obtain $\dfrac{RB - BN}{BN} = \dfrac{BN - BS}{BS}$. But

RB − BN = RA = AT. Why? Also, since BN = AB, the left

member of the proportion, $\dfrac{RB - BN}{BN}$, is equal to $\dfrac{AT}{AB}$.

Also, BN − BS = AB − AT = TB. Since BS = AT, the right member of the proportion, $\dfrac{BN - BS}{BS}$, is equal to $\dfrac{TB}{AT}$. Hence, $\dfrac{AT}{AB} = \dfrac{TB}{AT}$. Since it is perfectly all right to invert both members in a simple proportion (as will be proved in Section 4), we have $\dfrac{AB}{AT} = \dfrac{AT}{TB}$. Hence, our construction procedure for dividing a line segment into extreme and mean ratio is correct.

Now that you know what the Golden Section is and how to perform its construction for any given segment, you might enjoy making a mechanical device to enable you to quickly divide a segment into extreme and mean ratio. The device is illustrated in Figure 3.

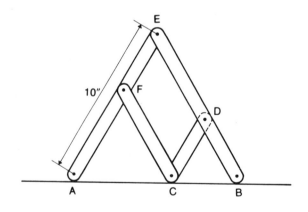

Figure 3

This device can be made out of wood, metal, or plastic. Cardboard would work, but it is not very durable. The strips AE and BE are cut the same length. Any length would do, but 10 inches is a convenient measure. These measurements are made from center to center of the small holes (each $\frac{1}{8}$ inch) drilled at the six labeled points. AF = FC and CD = DB. The strips are fastened together by small bolts at points F, C, D, and E with a wing nut being used at point F. Points F and D are located so as to divide strips AE and EB into extreme and mean ratio. In order to use the device, merely place the bolt holes A and B above the endpoints of a line segment and mark where hole C lies. Point C is the point of division.

PROBLEMS ???

1 Draw a segment about two inches long on a piece of paper. Divide the segment into extreme and mean ratio using a compass and straightedge. Repeat the construction for enough other segments of various lengths to feel sure that you really can perform the construction with ease.

2 Define what is meant by the Divine Proportion.

*3 The proof on page 7 is written in what mathematicians call *paragraph form*. Many times in writing a proof in paragraph form, obvious steps,as well as some of the reasons for each step,are omitted. Rewrite this proof in the more familiar two two-column form, which you have used in geometry. Be certain to justify each statement with an appropriate reason.

**4 Using a compass and straightedge only, how can you construct a segment whose length is the mean proportional of 1 and 5? That is, how can you construct a segment whose length is $\sqrt{5}$? Generalize your results. That is, describe how you would construct a segment whose length is the mean proportional of say m and n, where m and n are any two positive real numbers. (*Hint:* Clues can be found by examining Figure 2 and by studying Theorem 2.)

_____ SECTION 3

The Regular Pentagon
And Its
Relationship
To
The Golden Section

A *regular pentagon* is a five-sided polygon such that all of its sides are congruent and all of its interior angles are congruent. The regular pentagon is very closely related to the Golden Section. In this section and in Sections 5 and 6 we will examine some of the relationships between the Golden Section and the regular pentagon.

To begin, we need a regular pentagon. How do we go about constructing such a figure? A simple way to do this is to draw a circle; then using a protractor, measure five 72° angles at the center. Next, draw the five segments from the center to the circle and connect the five endpoints, as illustrated in Figure 4.

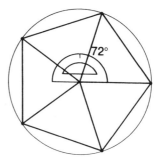

Figure 4

This sort of procedure, which makes use of a protractor, is perfectly logical if you think of the problem as an exercise in draftsmanship. However, in order to make our study of the construction of a regular pentagon more interesting and mathematically fruitful, let us rule out the use of a protractor. Instead, we will limit ourselves to the use of a compass and an unmarked straightedge. According to Hlavaty, Plato restricted his geometers to the use of a compass and an unmarked straightedge when considering various construction problems. However, there is nothing sacrosanct about restricting oneself to these two instruments when performing geometric constructions. In the past, mathematicians have also considered the consequences of restricting themselves to *only* a compass (geometric constructions done with a compass only are called *Mascheroni Constructions*) or *only* a straightedge when

doing constructions. Examine the list of references at the end
of the book for further information on these kinds of constructions.

As it has turned out in the history of mathematics, however,
the restriction to compass and unmarked straightedge when
doing constructions has led mathematicians to the considera-
tion--and eventual solution--of some very interesting constru-
tion problems. The most famous of these concern the trisection
of an angle, the duplication of a cube, and the squaring of a
circle. For further reading on these topics, see Tietze (47).

Construction problems done in the true Euclidean sense
would not allow us to transfer distances. This fact stemmed from
thinking of a compass as a collapsible one. This means that
the instant you removed either leg from the paper, the compass
would snap shut. Modern compasses, however, do not do this.
Even though modern compasses may seem more powerful
because of the transfer capability, it can be proved that every-
thing that can be done with a modern compass can be done
with a collapsible one. A proof of this fact can be found in
Eves (21). So the two instruments are mathematically equivalent.

If the use of a protractor is ruled out let us see how to
construct a regular pentagon with a compass and straightedge.
Two such constructions of the regular pentagon will be presented. The first method will produce the figure directly, while
the second method involves constructing a *regular decagon*
(a ten-sided polygon) and then joining alternate vertices.

To begin the first construction, draw a circle with center K
and radius KG. Construct $\overline{KT}$ perpendicular to $\overline{KG}$. Locate the
midpoint of $\overline{KT}$, call it M; now draw $\overline{MG}$. Bisect angle KMG,

calling Q the point of intersection of this bisector with $\overline{KG}$. At Q erect a perpendicular segment to $\overline{KG}$, calling P its point of intersection with the circle. The segment GP will be a side of the desired pentagon, and the other four sides can be struck off as shown in Figure 5.

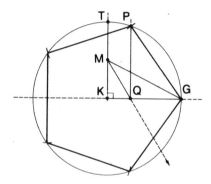

Figure 5

One proof of this construction is rather sophisticated and involves some clever manipulation of trigonometric identities. Refer to Appendix C for further information on this proof.

Now let us examine the second method for constructing a regular pentagon. This method points out clearly one way in which the Golden Section is related to the regular pentagon. To begin the construction, you will need to refer to Section 2, where you learned a method of dividing a segment $\overline{AB}$ into extreme and mean ratio. In fact, it will be convenient to use Figure 2 in this construction. Study Figure 2 on page 6 . All we need to do in order to get our pentagon is to draw a circle having its radius OA (See Figure 6) the same length as $\overline{AB}$ of Figure 2. Then on the circumference of this circle, mark off $\overline{AT}$

as a chord ten times. Connect in sequence the ten points, A, B, C, . . . , on the circle. This produces a regular decagon. To get a regular pentagon, merely connect alternate vertices.

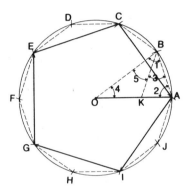

Figure 6

How can we convince ourselves that such a procedure will really give us a regular pentagon? The obvious answer, which has already been discussed, is to write out a mathematical proof. Where should we start?

One possible approach is the following. From your work in geometry, remember that in a regular n-gon the measure of each central angle is $\frac{360}{n}$. Therefore, if we could show that each of the five central angles of ACEGI has a measure of $\frac{360}{5}$, or 72, then we could be certain that we have a regular pentagon.

Taking this one step further, if we could only show that $m \angle BOA = 36$, then we would know we have a regular pentagon. Why?

To prove $m \sphericalangle \text{BOA} = 36$, we will use the following two theorems:

Theorem 3: If two sides of one triangle are proportional to two corresponding sides of a second triangle, and if the angles included by these sides are congruent, then the triangles are similar.

Theorem 4: The base angles of an isosceles triangle are congruent.

With these theorems in mind, let us look again at Figure 6. Draw in $\overline{\text{OB}}$. Using O as an endpoint, locate a point K on $\overline{\text{OA}}$ such that OK = BA. Draw $\overline{\text{BK}}$. Now if you think carefully, you can see that $\dfrac{\text{OB}}{\text{AB}} = \dfrac{\text{AB}}{\text{AK}}$ because of the way in which the figure was constructed and because of the Golden Section. Now OA = OB, so from Theorem 4 we know that $m \sphericalangle 1 = m \sphericalangle 2$. Hence from Theorem 3, $\triangle \text{OBA} \sim \triangle \text{BAK}$. So $m \sphericalangle 3 = m \sphericalangle 4$, since the corresponding angles of similar triangles are congruent. Now we can prove that $\triangle \text{OKB}$ is isosceles, so $m \sphericalangle 4 = m \sphericalangle 5$. Since $m \sphericalangle 3 = m \sphericalangle 4$, it follows that $m \sphericalangle 5 = m \sphericalangle 3$. With a little thought, we can convince ourselves that the sum of the measures of the angles of $\triangle \text{OBA}$ is equal to $5 \left(m \sphericalangle 4 \right)$. Since the sum of the measures of the angles of a triangle is 180, we have that $5 \left(m \sphericalangle 4 \right) = 180$. Hence, $m \sphericalangle 4 = 36$, and we have our desired result.

The preceding discussion of the proof of our second construction contains a number of missing parts. You will be asked to supply these "missing links" in Problem 6.

PROBLEMS ???

1 Construct several pentagons, using both of the construction procedures presented in this section.

2 Give some examples of the pentagon in nature; for example, in flowers, leaves, and animals. (See Section 11.)

3 Use a protractor to construct a regular 3-gon, 4-gon, 5-gon, and 6-gon. What are the names given to these polygons?

4 Draw all the diagonals of the pentagon that you constructed in Problem 3. Remember that a diagonal is any segment connecting two nonconsecutive vertices of a polygon.(The polygon need not be regular.)How many are there?

  Do the same thing for a triangle, a square and a hexagon. Try to discover a formula for the number of diagonals of a polygon of n sides.

5 The title of this section refers to a relationship between the Golden Section and a regular pentagon. Explain the relationship.

**6 In the proof of the second construction in this section there were quite a number of "missing links." This problem pertains to them.

a) In the paragraph preceding Theorem 3, it was mentioned that if we could show that each of the five central angles of ACEGI has a measure of 72, then we could be certain we had a regular pentagon. Why?

b) In the same paragraph, why does showing that $m \angle BOA = 36$ guarantee that we have a regular pentagon?

c) In the paragraph immediately following Theorem 4, prove that $\triangle OKB$ is isosceles.

d) Again in the same paragraph, prove that in $\triangle OBA$, $5\ (m \angle 4) = 180$.

PROJECTS ✳ ✳ ✳ ✳ ✳ ✳ ✳ ✳ ✳ ✳ ✳ ✳ ✳ ✳ ✳ ✳ ✳ ✳ ✳

1 Try this quick way to make a paper model of a pentagon as shown in the diagram. Just take a strip of paper 1 inch to $1\frac{1}{2}$ inches wide and about 12 inches long; then very carefully tie a knot in it and press it flat.

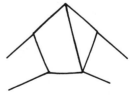

2 Illustrated below are a *cube* and a *tetrahedron*. These three-dimensional objects are just two of an infinite number of solids generally referred to as *polyhedrons*.

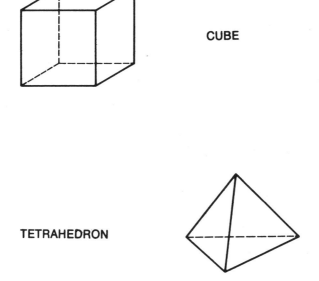

CUBE

TETRAHEDRON

The regular pentagon occurs among the faces of certain polyhedrons. The simplest such polyhedron is called the *dodecahedron,* which is one of the five so-called Platonic polyhedrons. A plan for constructing this solid is given on page 20. Try making one. You will need some poster board, a single-edge razor blade, some glue, and some cellophane tape. Paint can be used for decorative purposes. Look

The Regular Pentagon and Its Relationship
To The Golden Section

at the sketch. The dotted lines indicate that you are to score the cardboard; that is, only cut part-way through. This is so you will get sharp creases when you start folding the model together.

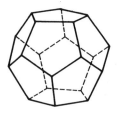

After you have cut out the model, just close it up in a regular three-dimensional figure. When you are through, your model should look like this.

Refer to Cundy and Rollet (19) for detailed plans for making many such polyhedrons.

SECTION 4

The Algebraic Representation
Of
The Golden Section

The preceding sections have dealt with some of the geometric aspects of the Golden Section. In Section 2 we learned
how to construct the Golden Section, that is, how to divide a
segment into extreme and mean ratio. In Section 3 we saw
how to construct a regular pentagon; also, we learned something of its relationship to the Golden Section. To pursue this
relationship more deeply, we will need to learn something of
the algebraic representation of the Golden Section.

Just what is meant by the title of this section?

Recall that in Section 1 we said that the Golden Section referred to the division of a segment $\overline{AB}$ by a point C such that
the two ratios $\dfrac{AB}{AC}$ and $\dfrac{AC}{CB}$ were equal. Now we will determine
the numerical values of these two ratios. This numerical value
is what we will call the algebraic representation of the Golden
Section.

To determine the value of these ratios, consider a segment
AB where C is the point that divides $\overline{AB}$ into extreme and mean
ratio (See Figure 7a). Let the lengths of $\overline{AB}$, $\overline{AC}$, and $\overline{CB}$ be
AB, AC, and AB − AC, respectively, so that $\dfrac{AB}{AC} = \dfrac{AC}{AB - AC}$.

Since we are interested only in the ratio of AB to AC, we could
let AB be any positive real number, substitute this number into
$\dfrac{AB}{AC} = \dfrac{AC}{AB - AC}$, and solve the resulting quadratic equation
for AC. Knowing both AB and AC, we could easily calculate
the desired ratio.

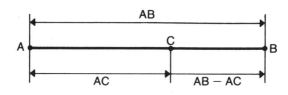

Figure 7a

As so often occurs in mathematics, however, the algebraic
manipulations can be simplified if we let the length of $\overline{AB}$
be 1. Upon making this substitution, the above description will

reduce to the following.

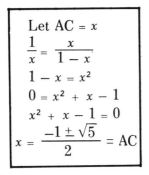

Let AC $= x$

$$\frac{1}{x} = \frac{x}{1-x}$$

$1 - x = x^2$

$0 = x^2 + x - 1$

$x^2 + x - 1 = 0$

$$x = \frac{-1 \pm \sqrt{5}}{2} = AC$$

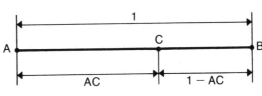

Figure 7b

The equation $x^2 + x - 1 = 0$, or $AC^2 + AC - 1 = 0$, algebraically characterizes the arithmetic of the Golden Section, so it is called the *characteristic equation.* The second solution to the characteristic equation (i.e., $\frac{-1 - \sqrt{5}}{2}$) is discarded, because it is negative and cannot represent the length of a line segment.

Let us examine Figure 7b in order to get clearly in mind just what the number $\frac{-1 + \sqrt{5}}{2}$ represents. If we agree that AC can be written as $\frac{AC}{1}$, then we see that AC represents the numerical value of the ratio $\frac{AC}{AB}$. In other words, if a segment is divided into extreme and mean ratio, then the ratio of the longer segment to the whole segment is *always* the number $\frac{-1 + \sqrt{5}}{2}$. Since $\frac{AB}{AC} = \frac{AC}{CB}$ by definition of the Golden Section, we know that $\frac{AC}{AB} = \frac{CB}{AC}$ (See Section 4, Problem 1). Hence,

$\dfrac{-1 + \sqrt{5}}{2}$ is also the ratio of the shorter segment to the longer.

This number is given a special name, which is $\dfrac{1}{\tau}$. (τ is a letter of the Greek alphabet and is named *tau*.) Hence, $\dfrac{1}{\tau} = \dfrac{-1 + \sqrt{5}}{2}$.

Let us carry this discussion one step further. If $\dfrac{1}{\tau} = \dfrac{-1 + \sqrt{5}}{2}$, then we know by proportion inverse (Section 4, Problem 1 again), that $\tau = \dfrac{2}{-1 + \sqrt{5}}$. Multiplying the right side of this equation by 1, which we conveniently express as $\dfrac{-1 - \sqrt{5}}{-1 - \sqrt{5}}$, we see upon simplifying our product that $\tau = \dfrac{1 + \sqrt{5}}{2}$.

Recalling our earlier work in this section, we know that

$$\frac{AC}{AB} = \frac{CB}{AC} = \frac{1}{\tau} = \frac{-1 + \sqrt{5}}{2}.$$

Hence, it follows that

$$\frac{AB}{AC} = \frac{AC}{CB} = \tau = \frac{1 + \sqrt{5}}{2}.$$

Why? Consequently, we see that the number τ, which equals $\dfrac{1 + \sqrt{5}}{2}$, is *always* the ratio of the whole segment to the longer segment, which is the same as the ratio of the longer segment to the shorter segment when a segment is divided into extreme and mean ratio. Not only have we determined the value of the ratios $\dfrac{AB}{AC}$ and $\dfrac{AC}{CB}$; but, in the process, we have also determined the value of their reciprocals, $\dfrac{AC}{AB}$ and $\dfrac{CB}{AC}$.

A summary of the material in this section is given below.

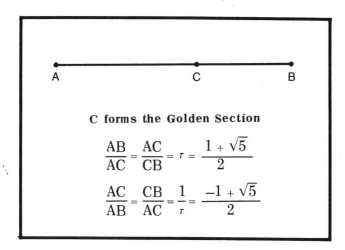

C forms the Golden Section

$$\frac{AB}{AC} = \frac{AC}{CB} = \tau = \frac{1 + \sqrt{5}}{2}$$

$$\frac{AC}{AB} = \frac{CB}{AC} = \frac{1}{\tau} = \frac{-1 + \sqrt{5}}{2}$$

PROBLEMS ???

1 Prove that the following properties are true for all real numbers a, b, c, d, where none of a, b, c, or d is zero.

a) If $\dfrac{a}{b} = \dfrac{c}{d}$, then $\dfrac{b}{a} = \dfrac{d}{c}$.

b) If $\dfrac{a}{b} = \dfrac{c}{d}$, then $\dfrac{a + b}{b} = \dfrac{c + d}{d}$.

c) If $\dfrac{a}{b} = \dfrac{c}{d}$, then $\dfrac{a}{c} = \dfrac{b}{d}$.

d) If $\dfrac{a}{b} = \dfrac{c}{d}$, then $\dfrac{d}{b} = \dfrac{c}{a}$.

e) If $\dfrac{a}{b} = \dfrac{c}{d}$, then $\dfrac{a - b}{b} = \dfrac{c - d}{d}$.

*2 Show that if you let AB = 3, then the ratio of $\dfrac{AC}{AB}$ will still

be $\dfrac{-1 + \sqrt{5}}{2}$ (See Figure 7a). Generalize your result. That

is, show that if you let AB be any positive number, then

the ratio $\dfrac{AC}{AB}$ will still equal $-1 + \sqrt{5}$.

3 Justify all of the steps in deriving the characteristic
equation beside Figure 7b.

4 We have mentioned that the characteristic equation referred
to in Figure 7b is called a quadratic equation. There are
generally three ways of solving quadratic equations:
a) factoring
b) completing the square
c) using the quadratic formula
Since $x^2 + x - 1 = 0$ cannot be factored over the integers,
solve the equation by the other two methods, b and c.

5 Besides being reciprocals of one another, τ and $\dfrac{1}{\tau}$ are related

in another way. To find what this relationship is, begin by

subtracting 1 from τ. (*Hint:* Express 1 as $\dfrac{2}{2}$.) State this

relationship.

6 Using a table of square roots, verify that approximations

to τ and $\dfrac{1}{\tau}$ are 1.618 and 0.618, respectively.

*7 On page 9 instructions are given for making a device
(illustrated in Figure 3) that could be used to divide a line
segment into the Golden Section. Prove that this device
really does what it is supposed to do. (*Hint:* Prove
$\triangle ABE \sim \triangle ACF$ and $\triangle ACF \sim \triangle CBD$.)

_____ *SECTION* 5

More On The Pentagon's Relationship To Tau

In Section 3 we saw how one construction of a regular pentagon was related to the geometric interpretation of the Golden Section. In this section we will examine how the pentagon can be related to the algebraic representation of the Golden Section. To make our study of this section more meaningful, several theorems are listed here.

More On The Pentagon's Relationship To Tau

Theorem 5: The measure of an angle inscribed in a circle is equal to one-half the measure of the intercepted arc.

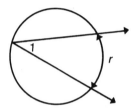

$$m \sphericalangle 1 = \frac{1}{2} \cdot r$$

Theorem 6: In a circle, equal arcs have equal chords.

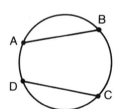

If $m \overset{\frown}{AB} = m \overset{\frown}{CD}$, then AB = CD.

Theorem 7: In a circle, equal chords have equal arcs.

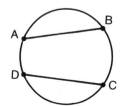

If AB = CD, then $m \overset{\frown}{AB} = m \overset{\frown}{CD}$.

With these theorems in mind, let us examine Figure 8 in order to learn more about the pentagon's relationship to the number τ.

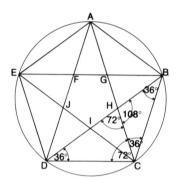

Figure 8

ABCDE in Figure 8 is a regular pentagon. Using Theorem 7, we can establish that $m\ \overset{\frown}{AB} = m\ \overset{\frown}{BC} = m\ \overset{\frown}{CD} = m\ \overset{\frown}{DE} = m\ \overset{\frown}{EA}$. By knowing that these arcs are equal, we can utilize Theorem 6 to prove that all of the diagonals of ABCDE have the same length.

Using Theorem 5, we can show that

$$m\ \measuredangle DBC = m\ \measuredangle ACB = m\ \measuredangle BDC = 36.$$

From geometry we know that when two angles of one triangle are congruent to two corresponding angles of a second triangle, the triangles are similar; therefore, $\triangle BCH \sim \triangle BDC$. Hence, $\dfrac{BH}{BC} = \dfrac{BC}{BD}$. As was done in Section 4, we let DB = 1 in order to simplify future computation. Then by substituting, we can write $\dfrac{BH}{BC} = \dfrac{BC}{1}$, which is equivalent to $BC^2 = BH$.

In Figure 8 we see that $m \sphericalangle DCA = 72$, $m \sphericalangle BDC = 36$, and $m \sphericalangle DHC = 72$. Hence, $\triangle DCH$ is isosceles and DC = DH = BC. From this fact and the fact that DB = DH + HB, we get $1 = BC + BH$ or $BH = 1 - BC$. Substituting this value into the equation $BC^2 = BH$, we obtain $BC^2 = 1 - BC$, or $BC^2 + BC - 1 = 0$.

Does this equation look familiar? Compare it to the one in Figure 7b. It is the characteristic equation again. We already know the solution to this equation. It is $BC = \dfrac{-1 + \sqrt{5}}{2} = \dfrac{1}{\tau}$.

Just what does this show? If we look again at Figure 8 (and remember that we let DB = 1), we see that we have shown that

$BC = \dfrac{BC}{1} = \dfrac{BC}{DB} = \dfrac{-1 + \sqrt{5}}{2} = \dfrac{1}{\tau}$. More simply, *the ratio of the*

length of any side of a regular pentagon to the length of any of

its diagonals is $\dfrac{1}{\tau}$. We should notice, of course, that by the

property of proportion inversion, we can also say that $\dfrac{DB}{BC} = \tau$.

That is, *the ratio of the length of any diagonal of a regular*

pentagon to the length of any of its sides is given by $\tau = \dfrac{1 + \sqrt{5}}{2}$.

PROBLEMS ??

1 Prove Theorem 5. (*Hint:* There are three cases to consider.)

2 Prove Theorem 6.

3 Prove Theorem 7.

4 Show that in Figure 8 all of the diagonals of pentagon ABCDE are equal in length.

*5 Prove that FGHIJ in Figure 8 is a regular pentagon.

6 Theorem 5 can be used to determine the measure of any interior angle of a regular n-gon. There is a formula, however, that enables you to determine this measure knowing nothing but n, the number of sides of the polygon. Can you discover this formula? (*Hint:* Select any vertex of the n-gon and draw all of the diagonals from this vertex only.) Does this formula apply to polygons that are not regular?

7 Your answer to Problem 6 probably includes a formula for finding the sum of the measures of the interior angles of a regular n-gon. What is this formula? Is this formula applicable to polygons that are not regular?

*8 In the proof that the ratio of any side of a regular pentagon to its diagonal is $\dfrac{-1 + \sqrt{5}}{2}$, we assumed that if two angles of a triangle are congruent, then the triangle is isosceles. Prove that this assumption is valid.

PROJECTS ✳ ✳ ✳ ✳ ✳ ✳ ✳ ✳ ✳ ✳ ✳ ✳ ✳ ✳ ✳ ✳ ✳ ✳ ✳

Construct five pentagons of equal size on a piece of paper.
Cut out all five pentagons. Divide two of the pentagons into
nine *isosceles* triangles, as shown below. Cut out these pieces.
Using the remaining three pentagons and the nine isosceles
triangles, form a single larger pentagon.

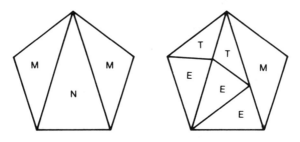

See Holt (35) for an analysis of the dimensions and areas of
these various triangles and pentagons.

Progressions,
Pentagons,
And
Pentagrams

In Section 5 we learned of the relationship that exists between the lengths of a diagonal and a side of any regular pentagon. In this section we will discover the relationship that exists between the lengths of the segments that form Figure 9. In order to do this, we need to digress for a moment to study briefly a topic from algebra.

We will begin this digression by examining the following pattern of numbers:

$$1, 2, 4, 8, 16, 32, 64$$

Notice that each number after 1 is found by multiplying the preceding number by 2. Thus, $2 = 1 \cdot 2$, $4 = 2 \cdot 2$, $8 = 4 \cdot 2$, and so on. Such a pattern of numbers that is formed by multiplying each preceding term by a constant is called a *geometric progression*. The constant is called the *common ratio* of the progression. In the example, the common ratio is 2.

In general, a geometric progression has the following form:

$$a, ar, ar^2, ar^3, ..., ar^{n-1} \text{ where}$$

a represents the first term of the progression and r is the common ratio. The number of terms in the progression is n, In order to determine a geometric progression, therefore, you need to know a, r, and n. For example, if $a = 3$, $r = \sqrt{2}$, and $n = 5$, then the geometric progression determined by these values of a, r, and n is:

$$3, 3\sqrt{2}, 3(\sqrt{2})^2, 3(\sqrt{2})^3, 3(\sqrt{2})^4,$$

or

$$3, 3\sqrt{2}, 6, 6\sqrt{2}, 12.$$

An important observation concerning geometric progressions is that the ratio of any two consecutive terms of the progression is the same as the ratio of any other two consecutive terms considered in the same order. To illustrate what is meant by this statement, we note in the progression 1, 2, 4, 8, 16, 32, 64, that $\frac{1}{2} = \frac{2}{4} = \frac{4}{8} = \frac{8}{16} = \frac{16}{32} = \frac{32}{64}$ and also that $\frac{2}{1} = \frac{4}{2} = \frac{8}{4} = \frac{16}{8} = \frac{32}{16} = \frac{64}{32}$. This observation will be of much importance in the development that follows. Thus ends the digression.

Let us look carefully at Figure 9. ABCDE is a regular pentagon, and the star-shaped figure formed by its diagonals is called a *pentagram*. Notice that the diagonals of ABCDE form not only a pentagram but also another smaller pentagon FGHIJ. Drawing the diagonals of this pentagon will form another pentagram. We could continue forming pentagons and pentagrams in this manner for as long as we wished, but we will restrict our discussion to the two pentagons and two pentagrams illustrated.

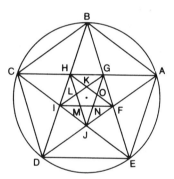

Figure 9

It can be proved that in Figure 9 there are six different lengths of line segments in the two pentagons and the two pentagrams. (We will accept this without proof.) For convenience, we will denote the length of these segments by lower case letters as follows:

BE = *a*, DE = *b*, FE = *c*, JF = *d*, OF = *e,* and ON = *f.*

Before going on, be sure you understand that all other segments in the figure could be put into one of these six categories.

Using the angle measures discovered in Section 5(See Figure 8), it is not difficult to show that △BED ~ △DEF. It certainly follows that $\frac{a}{b} = \frac{b}{c}$. It is also not difficult to establish

that $\triangle DEF \sim \triangle EJF$. Therefore, $\frac{b}{c} = \frac{c}{d}$. *(In Problems 6 and 7 you

will be asked to prove that $\triangle EJF \cong \triangle HJF$ and that FGHIJ is a

regular pentagon.)* Assuming the two statements in parentheses

to be true, we may say that HF = FE = c. Since $\triangle HFJ \sim \triangle JFO$,

it follows that $\frac{HF}{JF} = \frac{JF}{OF}$, or that $\frac{c}{d} = \frac{d}{e}$. Finally, we can say that

$\triangle JFO \sim \triangle FNO$, and hence, $\frac{d}{e} = \frac{e}{f}$. Chaining all of these propor-

tions together, we get the following:

$$(1) \quad \frac{a}{b} = \frac{b}{c} = \frac{c}{d} = \frac{d}{e} = \frac{e}{f}.$$

Now just what does this chain of equalities tell us?
To get the answer to this question, consider the following
progression of segment lengths:

$$(2) \quad a, \ b, \ c, \ d, \ e, \ f.$$

Since (1) is true, then (2) must be a geometric progression. Why?
From what we learned about geometric progressions, this means
that b, c, d, e, and f can each be obtained from a by multiplying
a by a power of r, the common ratio of the progression. In other
words, $b = ar$, $c = ar^2$, $d = ar^3$, $e = ar^4$ and $f = ar^5$. Thus (2)
becomes:

$$a, \ ar, \ ar^2, \ ar^3, \ ar^4, \ ar^5.$$

Are you curious as to what the value of r is? Could you make a guess? One way to find r would be to notice in Figure 9 that BE = BF + FE, or since BF = DE, that BE = DE + FE. That is, $a = b + c$ or $a = ar + ar^2$. Simplifying, we have $1 = r + r^2$ or $r^2 + r - 1 = 0$. But this is our old friend, the characteristic equation again (See Figure 7b)! Since its solution is

$$r = \frac{-1 + \sqrt{5}}{2} = \frac{1}{r},$$

we see that each term in progression (2) can be obtained by multiplying the preceding term by $\frac{1}{r}$.

In the progression a, b, c, d, e, f the elements are in order of decreasing magnitude. That is, the first element represents the length of the longest segment and the last element represents the length of the shortest segment. Therefore, what we have just shown is that in Figure 9 each segment's length can be obtained from the next longer segment's length by multiplying the latter by the number $\frac{1}{r}$. In Problem 9 we will be asked to show that each segment's length can be obtained from the next shorter segment's length by multiplying the latter by r. Thus, we have found in this section a very interesting relationship among the lengths of the segments in Figure 9.

If we were to continue to inscribe pentagrams in Figure 9, we could continue finding the lengths of the successively shorter segments that would be produced. Continuing to find these smaller and smaller lengths forever would create what is called an infinite geometric progression. Problem 15 relates to such a progression.

PROBLEMS ??

1 Construct a geometric progression containing seven elements for each of the following, where a is the first term and r is the common ratio.

a) $a = 1, r = 3$ c) $a = \dfrac{1}{5}, r = 5$

b) $a = \sqrt{3}, r = 2$ d) $a = .1, r = .01$

2 Tell whether each progression is geometric. If the progression is geometric, give the value of r.

a) $1, 2, 3, 4, 5$ c) $2, \sqrt{2}, 1, \dfrac{\sqrt{2}}{2}, \dfrac{1}{2}$

b) $\dfrac{1}{3}, \dfrac{1}{6}, \dfrac{1}{12}, \dfrac{1}{24}, \dfrac{1}{48}$ d) $6, 8.5, 11, 13.5, 16$

3 Use Figure 9. Prove that the five triangles DJE, EFA, AGB, BHC, and CID are congruent.

4 Use Figure 9. Prove that the five triangles EFJ, AGF, BHG, CIH, and DJI are congruent.

5 Use Figure 9. Prove that the ten triangles EAG, EDI, AEJ, ABH, BFA, BCI, CBG, CDJ, DCH, and DEF are congruent.

6 Use Figure 9. Prove that FGHIJ is a regular pentagon.

7 Use Figure 9. Prove that the ten triangles EFJ, AGF, BHG, CIH, DJI, HJF, IFG, JGH, FHI, and GIJ are congruent. Notice that you proved the first five triangles congruent in Problem 4.

8 If you examine Figure 9, it appears as though the pentagon ABCDE is formed by five overlapping *rhombi*. (A rhombus is a parallelogram with two adjacent sides congruent.) Can

you find these rhombi? Prove that they are really rhombi. Also in the figure are two smaller varieties of rhombi. Find them and prove that they are rhombi.

9 Show that the increasing pattern of segment lengths in Figure 9 forms a geometric progression whose common ratio is $\tau = \dfrac{1 + \sqrt{5}}{2}$. (*Hint:* Begin by rewriting the expression

$$\frac{a}{b} = \frac{b}{c} = \frac{c}{d} = \frac{d}{e} = \frac{e}{f} \ as \ \frac{f}{e} = \frac{e}{d} = \frac{d}{c} = \frac{c}{b} = \frac{b}{a}.)$$

*10 The figures below are regular pentagons of the same size. Show that the areas of the shaded regions of the three pentagons form a geometric progression with common ratio $\dfrac{1}{\tau}$. (*Hint:* The shaded area of Pentagon I can be thought of as five times the area of $\triangle ABF$. The shaded area of Pentagon II can be thought of as five times the area of $\triangle ABG$. The shaded area of Pentagon III can be thought of as five times the area of $\triangle AFG$. What is true of the bases of these triangular regions?)

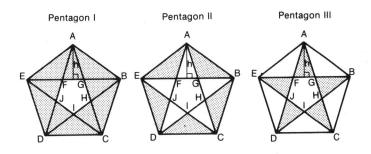

Pentagon I   Pentagon II   Pentagon III

11 Pictured below is a portion of the emblem of The Fibonacci Association. If the length of CD is 1, determine the lengths of $\overline{DE}$, $\overline{EF}$, $\overline{FG}$, $\overline{GH}$, $\overline{HI}$, $\overline{IJ}$, and $\overline{JK}$.

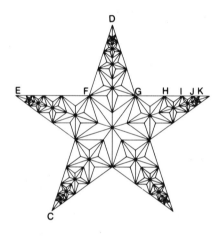

12 Suppose that in Figure 9, the length of $\overline{ON}$ is 1. Find the lengths of the remaining five classes of segments.

**13 In the drawing below, △MNO is an inscribed isosceles triangle whose vertex angle has a measure of 36 and whose base angles have a measure of 72. Using this drawing, how could you construct, using a compass and a straightedge only, a regular pentagon that would be inscribed in the circle? Prove that your construction procedure is correct.

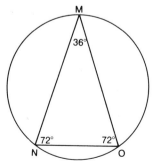

*14 The famous five-star insigne of the United States' top
military commanders is illustrated in the diagram below.
If the length of $\overline{AB}$ = 1, find the lengths of $\overline{CD}$, $\overline{EG}$, $\overline{EF}$, $\overline{FH}$,
and $\overline{HQ}$.

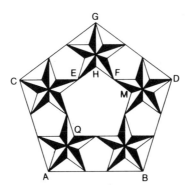

15 Suppose that in Figure 9 we let the length of $\overline{DB}$ be 2 inches.
Also, suppose that we imagine continuing to draw in the
other inscribed pentagrams. Using the formula $\dfrac{a}{1-r}$, where
a is the first term of a geometric progression and r is the
common ratio, determine what the sum of the lengths of all
of the different sized segments would be. Express your
answer to two decimal places.

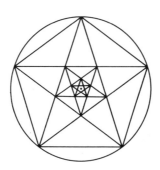

PROJECTS ✳ ✳ ✳ ✳ ✳ ✳ ✳ ✳ ✳ ✳ ✳ ✳ ✳ ✳ ✳ ✳ ✳ ✳ ✳

On a large piece of poster board (perhaps a colored piece), draw a circle. Then use a protractor to construct an inscribed, regular decagon. Draw in the five diameters of the circumscribed circle, connecting opposite vertices of the decagon. Next locate the ten pentagons that can be drawn inside the decagon, using each side of the decagon as one of the sides of the pentagons. Now draw all of the diagonals of each pentagon. There they are—ten pentagrams. Look carefully. Now you see them, now you do not! Distinguish these ten pentagrams through shading, coloring, outlining, or any other method that is applicable. See the diagram below for an idea of how to construct just one of

these ten pentagrams. (By the way, from your study of Sections 3 and 4, can you determine what the values of the

ratios $\dfrac{AC}{AB}$ and $\dfrac{AB}{AC}$ are?)

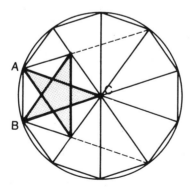

Golden
Rectangles

In this section we will discover what *Golden Rectangles* are. In addition we will study some properties of this very special figure. In Section 7 much information regarding Golden Rectangles is to be found in the problems, especially Problems 1, 2, 5–11, 14, and 15.

To begin, let us examine the rectangles in Figure 10.

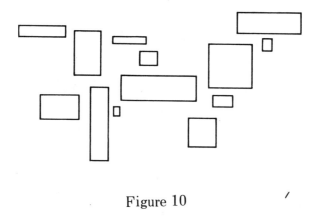

Figure 10

If we had to select the most "beautiful" of these rectangles, which would it be? In fact, of all the rectangles that exist in the world, which are the most beautiful or pleasing? These questions may sound silly. However, the early Greeks not only took such questions seriously, but they actually suggested an answer. Their answer is directly related to the Golden Section.

Remember that in Section 2 we learned how to divide a line segment into extreme and mean ratio. Using the procedure described in Section 2 (See Figure 11), suppose we locate the point C on $\overline{AB}$ so that C forms the Golden Section. Now if we use the two segments $\overline{AC}$ and $\overline{CB}$ as the sides of a rectangle, we have formed what is called a Golden Rectangle. ACFM in Figure 11 is a Golden Rectangle. A rectangle so formed was considered by many early mathematicians and artists to have, in some sense, the most pleasing proportions. They were thus considered to be the most beautiful of all rectangles. Much evidence of the conscious use of the proportions of Golden Rectangles can be found in early Greek art and architecture.

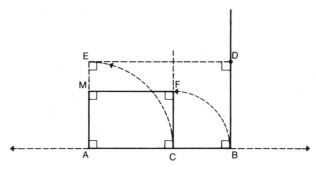

Figure 11

In the remainder of this section and in the problems that follow, we will discover some of the properties of Golden Rectangles.

The first and perhaps most obvious property can be found from a comparison of Figures 7b and 11. Examination of these figures tells us that the ratio of the longer side to the shorter side of a Golden Rectangle is $\tau = \dfrac{1 + \sqrt{5}}{2}$. What is the ratio of the shorter side to the longer? From what has been said in this paragraph, we see that another Golden Rectangle can be drawn in Figure 11. That is, we could let $\overline{AB}$ be one side of the "new" Golden Rectangle and $\overline{AC}$ the other. In Figure 11, therefore, ABDE is also a Golden Rectangle where AC = AE. It should be clear to the reader that by varying the length of $\overline{AB}$ we can construct Golden Rectangles of different sizes.

To find another property of Golden Rectangles that is not so obvious, consider Figure 12. AEFD is a Golden Rectangle, and ABCD is a unit square.

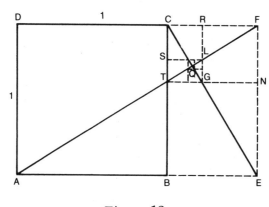

Figure 12

Draw diagonal $\overline{AF}$. Then construct a segment from point E perpendicular to $\overline{AF}$ at Q. This segment EQ will pass through point C (See Problem 9) . Rectangle CFEB is called the *reciprocal* of rectangle AEFD. (Note: The notion of *reciprocal rectangles* need not have anything to do with Golden Rectangles. In general, the reciprocal of a rectangle is another rectangle smaller in size but *similar* in shape to the original rectangle, with the property that one end of the original is a side of the reciprocal. It is also a general property of reciprocal rectangles that their diagonals are perpendicular. In fact, it is this property that enables us to construct very easily the reciprocal of any given rectangle. (See Problem 8.)

Once we have determined, using the above procedures, the reciprocal of the Golden Rectangle in Figure 12, we are in a position to form the reciprocal of the reciprocal rectangle, and so on, until our picture becomes too small to work with. To see how this is done, merely locate the point of intersection of $\overline{CB}$ and $\overline{AF}$; call it T. Construct $\overline{TN} \perp \overline{FE}$. Rectangle CFNT is the reciprocal of rectangle CBEF. Let $\overline{TN}$ and $\overline{CE}$ intersect at G and make $\overline{GR} \perp \overline{CF}$. TGRC is the reciprocal of TCFN. And so it goes (See Problem 14). In Problem 9 we will prove that successive reciprocal rectangles of Golden Rectangles form squares with their "parent" rectangles. Thus, ABCD, BENT, NFRG, and so on, are squares. We can see by examining Figure 13 why the Golden Rectangle is sometimes called the rectangle of the whirling squares.

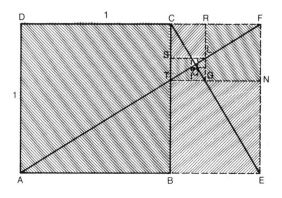

Figure 13

We note, then, that another property of the Golden Rectangle is that its successive reciprocal rectangles "cut off" squares that spiral in toward the point of intersection of one of its diagonals with a diagonal of the first reciprocal rectangle. We will encounter this spiraling again in a later section.

PROBLEMS??

1 Give a definition of a Golden Rectangle. (*Hint:* See the first property of Golden Rectangles mentioned in this section.)

2 Given the segment $\overline{GR}$ below, construct a Golden Rectangle whose perimeter is equal to two times the length of $\overline{GR}$. (*Hint:* Consult Figure 11.)

G   R

3 Find the lengths of $\overline{OB}$, $\overline{OC}$, $\overline{OD}$, $\overline{OE}$, $\overline{OF}$, and $\overline{OG}$ in the figure below.

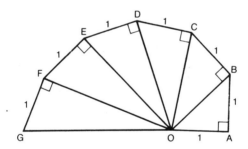

4 Find the lengths of $\overline{AE}$, $\overline{AF}$, $\overline{AG}$, $\overline{AH}$, $\overline{AI}$, and $\overline{AJ}$ in the figure below.

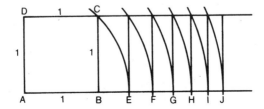

*5 Let ABCD be a Golden Rectangle; let ECBF be its
    reciprocal.

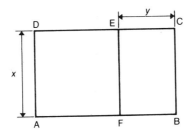

a) What are the values of the ratios $\dfrac{CB}{EC}$ and $\dfrac{DC}{DA}$? (Why?)

b) Verify that if the length of $\overline{DA}$ is x and the length of $\overline{EC}$
    is y then $\dfrac{x}{y} = \dfrac{x + y}{x}$.

c) Show that $\dfrac{x}{y} = 1 + \dfrac{y}{x}$.

d) Why does $\left(\dfrac{x}{y}\right)^2 = \dfrac{x}{y} + 1$?

e) Why does $\left(\dfrac{x}{y}\right)^2 - \dfrac{x}{y} - 1 = 0$? Does this equation look like
    something you have seen before in this booklet? (What?)

f) What is another name you have learned for $\dfrac{x}{y}$.

g) Give the solution of $\left(\dfrac{x}{y}\right)^2 - \dfrac{x}{y} - 1 = 0$ in terms of $\dfrac{x}{y}$.

h) Referring to the diagram, what does the ratio $\dfrac{x}{y}$ represent?

6 Let us find another property of the Golden Rectangle.

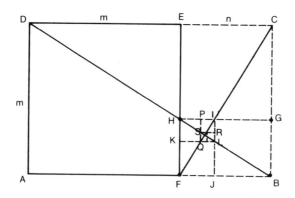

In the rectangle above, AD = *m* and EC = *n*. So
DC = DE + EC = *m* + *n*, since AD = DE. Starting with ABCD
and continuing with six successive reciprocal rectangles,
let us indicate the dimensions of each rectangle.

	ABCD	ECBF	FBGH	FJIH	HKLI	QLIP	RSQL
longer side	*m* + *n*	*m*	*n*	*m* − *n*	2*n* − *m*	2*m* − 3*n*	5*n* − 3*m*
shorter side	*m*	*n*	*m* − *n*	2*n* − *m*	2*m* − 3*n*	5*n* − 3*m*	5*m* − 8*n*

Can you find the rule of formation for the lengths of the
sides of the progressively smaller reciprocal rectangles?
Determine a pattern in the coefficients of *m* and *n*. More
will be said about this pattern later on.

*7 Now that you know that the ratio of the longer side to the shorter side of a Golden Rectangle is r and that the ratio of the shorter side to the longer side is $\frac{1}{r}$, go back to Figure 9 in Section 6 and see how many Golden Rectangles you can form using only the segments that you find in that figure. Verify your results. (*Hint*: You may want to review Section 6.)

8 Draw several rectangles on a sheet of paper. Construct the reciprocal of each of these.

**9 a) Since AEFD in Figure 12 is a Golden Rectangle and ABCD is a unit square, show that $\overline{AF}$, the diagonal of Golden Rectangle AEFD, and $\overline{CE}$ are perpendicular. Do not assume that CFEB is the reciprocal rectangle of AEFD, because that is in effect what you are trying to establish.

b) Consider the Golden Rectangle AEFD with diagonal $\overline{AF}$. Show that by drawing line EM perpendicular to $\overline{AF}$, intersecting $\overline{DF}$ at C, and then dropping $\overline{CB}$ perpendicular to $\overline{AE}$, a square ABCD will result.

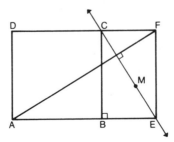

10 In Figure 12 show that TBEN, RGNF, RCSL, and so on are squares.

*11 Let a line segment AB have a length of one unit. Show how to construct a Golden Rectangle whose perimeter is $1 + \sqrt{5}$. (*Hint:* See Figure 11.)

**12 An isosceles triangle is called a *Golden Triangle* if the ratio of one of its sides to the base is $\tau = \dfrac{1 + \sqrt{5}}{2}$.

Triangle ABC below is a Golden Triangle.

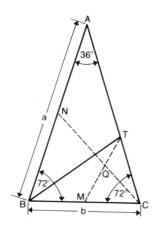

a) $\overline{BT}$ is the bisector of ∢ABC. How are △ABC and △BCT related? By what theorem in plane geometry do we know this?

b) What three segments are equal in length?

c) Let the length of $\overline{AB}$ be a and the length of $\overline{BC}$ be b. Show that

$$\frac{a}{b} = \frac{b}{a - b}.$$

d) From the equation in part c derive the equation
$\left(\dfrac{a}{b}\right)^2 - \dfrac{a}{b} - 1 = 0$. Does this equation look familiar?

e) What is the solution of $\left(\dfrac{a}{b}\right)^2 - \dfrac{a}{b} - 1 = 0$ in terms of $\dfrac{a}{b}$?

f) What does the ratio $\dfrac{a}{b}$ represent? From part e and the
definition of Golden Triangle, what kind of triangle do
you know to be a Golden Triangle?

g) How could you go about generating smaller and smaller
(or bigger and bigger, if you wish) isosceles triangles
similar to △ABC? Generate some of these triangles to
convince yourself that they spiral around point Q, which
is the point of intersection of medians $\overline{TM}$ and $\overline{CN}$.

h) Is it possible for a Golden Triangle to have angle
measures other than 36, 72, 72? (Why?)

**13 An alternative way to define a Golden Triangle is the
following: Suppose in the drawing below that △ACB ~ △BDC.
Then △ABC is a Golden Triangle if the ratio of the area of
△ABC to the area of △ADB is τ.

Show that if △ABC is a
Golden Triangle such that
△ACB ~ △BDC , then

$$\dfrac{\text{area of } \triangle ABC}{\text{area of } \triangle ADB} = \tau.$$

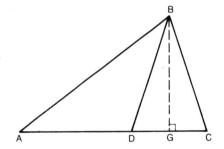

(*Hint:* Consider the altitude $\overline{BG}$.)

14 We have learned that the reciprocal of a rectangle is similar to the original rectangle. In the drawing below, rectangle GNBC is the reciprocal of rectangle ABCD. Prove that rectangles GNBC and ABCD are similar. (*Note:* Do not assume that ABCD is a Golden Rectangle.)

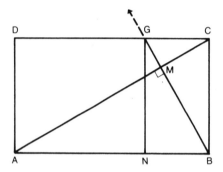

**15 Another topic that can be related to the Golden Rectangle and the Golden Triangle is that of the so-called *equiangular* or *logarithmic spiral*. We can get the "feel" of this spiral by first constructing a Golden Rectangle and several reciprocal rectangles as shown in the figure below.

Starting at A, draw $\overline{AE}$. $\overline{AE}$ forms an angle with a measure of 45 with $\overline{AB}$. (Why?) Now draw $\overline{EF}$. It makes an angle with a measure of 45 with $\overline{EC}$.

(Why?) Then draw $\overline{FG}$. What angle does it make with $\overline{FB}$? A continuation of this process yields a broken kind of spiral that tends toward the point of intersection of $\overline{DB}$ and $\overline{CL}$.

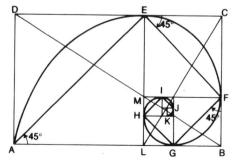

a) Show that $\left(\dfrac{1}{\tau}\right)^2 = 1 - \dfrac{1}{\tau}$.

b) Assuming that ALED is a unit square, show that the limit of the sum of the lengths AE, EF, FG, GH, HI, ... is $\tau^2\sqrt{2}$. (*Hint*: You may wish to use the formula for finding the sum of an infinite geometric progression, which can be found in Problem 15 of Section 6.)

c) What is a two-decimal approximation of $\tau^2\sqrt{2}$?

d) If your school has access to a computer facility, and if you or a friend know how to program, write a computer program that will print out the successive sums:

AE, AE + EF, AE + EF + FG, AE + EF + FG + GH, ...

Show that these sums in the computer output do approach or yield successively better approximations to your answer in part b.

e) Will the sum in part d ever equal $\tau^2\sqrt{2}$?

We can get a much closer approximation to the true logarithmic spiral by drawing successively smaller quarter-circles, as indicated in the diagram. Although this is not a true logarithmic spiral, the difference between this spiral and the real one is so slight on a drawing of this size that it will be an adequate approximation for our work.

It is interesting to note that the logarithmic spiral also passes through the vertices of the Golden Triangles in

the drawing below. You might wish to consult your school
or local library and try to find some additional information
on the logarithmic or equiangular spiral.

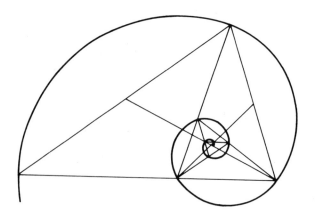

16 Figure ABCD below is a square, with E, F, G, and H
dividing the sides into extreme and mean ratio. Prove that
GHEF is a Golden Rectangle.

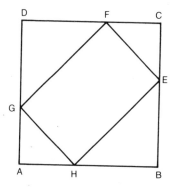

**17 Below are shown four regular geometric figures, each
inscribed in a circle. In each of these figures the length
of a side is related to the radius of the circumscribed circle.
That is, the length of a side can be given by an expression
involving R, the radius of the circumscribed circle.

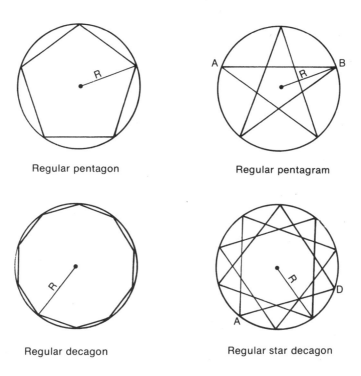

Regular pentagon   Regular pentagram

Regular decagon   Regular star decagon

In parts a through d, derive formulas for finding the lengths
of the sides of these figures in terms of R, the radius of the
circumscribed circle. (*Note:* The proofs of parts a, c, and d

depend upon the concept of the Golden Triangle which is discussed in Section 7, Problem 12.)

a) Prove that the length of the side of a regular pentagon is

$$\left(\frac{R}{2}\right)\sqrt{10 - 2\sqrt{5}}.$$

b) Prove that the length of the side $\overline{AB}$ of the pentagram is

$$\left(\frac{R}{2}\right)\sqrt{10 + 2\sqrt{5}}.$$

c) Prove that the length of the side of a regular decagon is

$$\frac{2R}{1 + \sqrt{5}} = \frac{R}{r}.$$

d) Prove that the length of the side $\overline{AD}$ of a *regular star decagon* is

$$\frac{2R}{\sqrt{5} - 1} = Rr.$$

PROJECTS ✳

As a project at the end of Section 3 a construction of one of the five Platonic solids was given. Below is a pattern for constructing another of these solids.

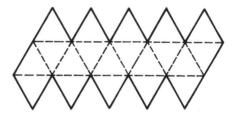

This solid is the most complex of the five, and is called an *icosahedron*. It has 20 equilateral triangular faces. The same directions for construction given in the project in Section 3 should be used here. The completed model is shown below.

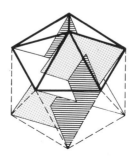

If we look carefully at this drawing of the completed model, we will notice that the five faces surrounding each vertex of the solid form a *regular pentagonal pyramid*. It can be proved that three mutually perpendicular rectangles can be placed inside the icosahedron as shown. Assuming that each edge of the polyhedron has length 1, determine the length of one of the longer sides of the rectangles (See Section 5). What kind of rectangles are these?

Put together one of these Platonic solids.

SECTION 8

Rabbits
And
Rectangles

An important sequence in mathematics that is neither geometric nor arithmetic was discovered more than 800 years ago
in connection with a problem about rabbits. According to
Sarton (42) and Eves (21), Leonardo Fibonacci , an Italian who
is considered by some scholars to be the greatest mathematician
of the Middle Ages, wrote a book in which the following question was raised: Suppose we have two newborn rabbits, one a
male and the other a female. How many pairs of rabbits will

there be at the end of one year if we assume that (1) rabbits begin to bear young two months after their own birth and (2) after reaching the age of two months each pair of rabbits produces another mixed pair every month thereafter?

Now we will solve this problem to discover for ourselves Fibonacci's sequence. Study Figure 14, which illustrates the rabbit population for the first seven months.

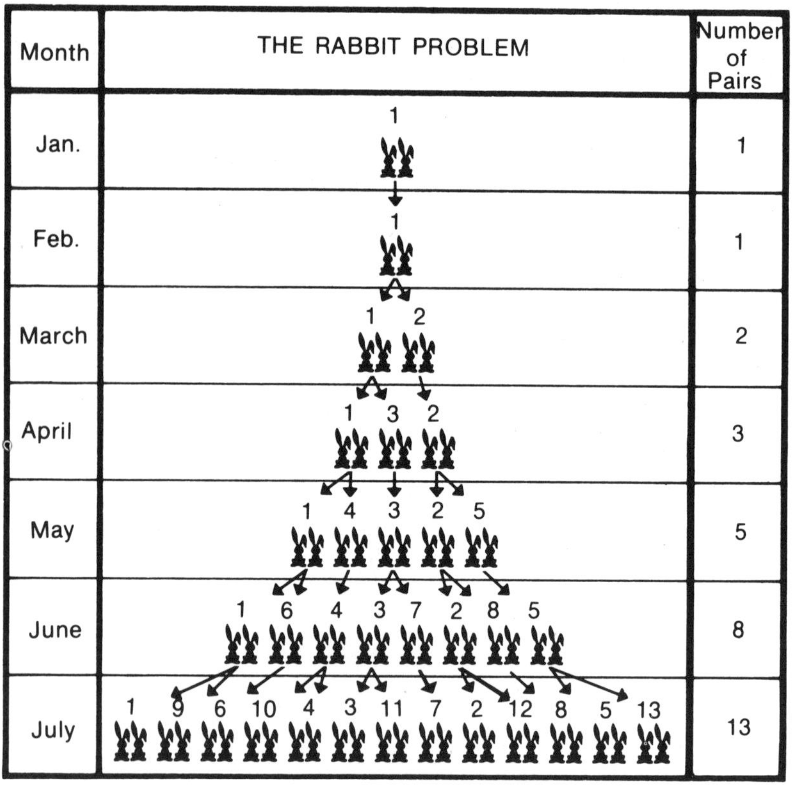

Month	THE RABBIT PROBLEM	Number of Pairs
Jan.	1	1
Feb.	1	1
March	1 2	2
April	1 3 2	3
May	1 4 3 2 5	5
June	1 6 4 3 7 2 8 5	8
July	1 9 6 10 4 3 11 7 2 12 8 5 13	13

Figure 14

Since we are concerned with rabbit growth for a period of one year, let us assume that our original pair of rabbits was born on January 1. Notice that at the end of February there is still only one pair. However, on March 1, our original pair, being two months old, produces a new pair (labeled "2"). On April 1 the original pair produces another new pair (labeled "3"). The pair labeled "2" produces no offspring, because they are only one month old. At the beginning of the fifth month (May) the original pair produces a fourth pair (labeled "4"); the third pair does not produce, because they are still too young; the second pair produces a fifth pair (labeled "5"). Therefore, at the end of five months we have five pairs of rabbits. Examine Figure 14 very carefully and verify that at the end of June and July we would have eight and thirteen pairs of rabbits, respectively. Now that we know the number of pairs of rabbits for each of the first seven months, we can write the first seven terms of the sequence as follows:

$$1, 1, 2, 3, 5, 8, 13, \ldots$$

Study the sequence carefully and try to observe a relationship between the terms. If you can discover a relationship, then it will be simple to determine the next few terms and obtain the answer to the problem.

In this case, the relationship that you should have observed is that each term, beginning with the third is the sum of the previous two terms. For example, the third term (2) is the sum of the first term (1) and the second term (1); the fourth term (3) is the sum of the second term (1) and the third term (2); and so on. If we were to continue our drawing in Figure 14, we would see that for any given month the number of pairs of rabbits is equal to the sum of the number of pairs of rabbits in the previous two months. This is the rule of formation that determines the *Fibonacci sequence*. In Problem 1, we will use this pattern to answer the famous rabbit problem.

Now let us discover how the Fibonacci sequence is related to Golden Rectangles. This relationship can be observed by examining the chart on page 50 which accompanies Problem 6 of Section 7. Turn back to that problem and notice the pattern being generated by the absolute values of the coefficents of m and n that are related to the longer and shorter sides of the successively smaller Golden Rectangles. Notice that in the row labeled "longer side," the absolute values of the "m's", beginning with rectangle FJIH, are 1, 1, 2 and 3. In the same row, the absolute values of the "n's," beginning with the rectangle FBGH, are 1, 1, 2, 3 and 5. Similarly for the row labeled "shorter side," beginning with the rectangle FBGH, the absolute values of the m-coefficients are 1, 1, 2, 3 and 5 while the absolute values of the n-coefficents, beginning with rectangle ECBF, are 1, 1, 2, 3, 5 and 8. Extrapolating from the chart, we can conclude that as additional dimensions are entered for progressively smaller reciprocal rectangles, the absolute values of the m's and n's associated with these dimensions will be successively larger Fibonacci numbers. Thus we see how the Fibonacci sequence can be related to Golden Rectangles.

The Fibonacci sequence is also related to the Golden Section in another very interesting way. Recall that the characteristic equation has the form $x^2 + x - 1 = 0$. An equivalent equation is $x = \dfrac{1}{1 + x}$. (How is this equation obtained?) Now

we make a clever move; in the right side of this equation we

replace x by $\dfrac{1}{1 + x}$. Doing this gives $x = \dfrac{1}{1 + \dfrac{1}{1 + x}}$; once

again replacing x in the right side of this equation by

$\dfrac{1}{1 + x}$, we get $x = \dfrac{1}{1 + \dfrac{1}{1 + \dfrac{1}{1 + x}}}$. If we imagine doing this

forever, we produce what is called a *continued fraction*. It looks like this:

$$x = \cfrac{1}{1 + \cfrac{1}{1 + \cfrac{1}{1 + \cfrac{1}{1 + \cfrac{1}{1 + \ldots}}}}}$$

At first glance the relationship of this fraction to the Fibonacci sequence is rather obscure. The relationship becomes clearer, however, if we rewrite the right side of the above expression as a sequence of fractions called *convergents*. Thus we have:

(I) $\dfrac{1}{1}$, $\dfrac{1}{1+\dfrac{1}{1}}$, $\dfrac{1}{1+\dfrac{1}{1+\dfrac{1}{1}}}$, $\dfrac{1}{1+\dfrac{1}{1+\dfrac{1}{1+\dfrac{1}{1}}}}$, $\dfrac{1}{1+\dfrac{1}{1+\dfrac{1}{1+\dfrac{1}{1+\dfrac{1}{1}}}}}$,

$\dfrac{1}{1+\dfrac{1}{1+\dfrac{1}{1+\dfrac{1}{1+\dfrac{1}{1}}}}}$, $\dfrac{1}{1+\dfrac{1}{1+\dfrac{1}{1+\dfrac{1}{1+\dfrac{1}{1+\dfrac{1}{1}}}}}}$, ...

For additional information on continued fractions and convergents, see Olds (40).

When we simplify these fractions, we obtain the following sequence:

(II) $\qquad 1, \frac{1}{2}, \frac{2}{3}, \frac{3}{5}, \frac{5}{8}, \frac{8}{13}, \frac{13}{21}, \ldots$

Now, the relationship of sequence (II) to the Fibonacci numbers is considerably more obvious. The numerators and denominators are the Fibonacci numbers with those in the denominators always one Fibonacci number ahead. But that is not all! Remember what x in the characteristic equation turned out to be? Numerically, $x = \dfrac{-1+\sqrt{5}}{2} = \dfrac{1}{\tau} \approx 0.61803$. Suppose we convert to decimals each of the fractions in sequence II. We get the following values, some of which are approximations.

(III) 1.00000, 0.50000, 0.66666, 0.60000, 0.62500, 0.61538, 0.61905, . . .

A careful examination of the decimals in sequence III reveals that they appear to be approaching the value $\frac{1}{\tau}$ as a limiting value. Although it has been proved that this is in fact the case, see Vorob'ev (49), we will accept without proof that the limit of sequence III is $\frac{1}{\tau}$. Can you see from this discussion how to use the Fibonacci sequence to get closer and closer approximations to the special number $\frac{1}{\tau}$? (See Problems 3 and 4 on page 68.)

Much has been written concerning the Fibonacci numbers. In fact, the Fibonacci Association exists in the United States. This organization publishes *The Fibonacci Quarterly*, a periodical containing articles relating to Fibonacci numbers. The list of references at the end of this booklet provides sources related to this sequence. Also, Section 9 contains a puzzle that can be closely related to the Fibonacci numbers.

PROBLEMS ??

1 Determine the answer to the rabbit problem posed by Fibonacci.

2 Expand sequence I in this section to include the first ten convergents. Then transform these convergents into fractions, as in sequence II. Finally, convert each fraction into its decimal equivalent correct to five decimal places as in sequence III. Carry the decimal conversions to five places.

3 From the results of Problem 2, explain how you can use the Fibonacci sequence to obtain closer and closer approximations to $\frac{1}{\tau}$.

4 Explain a way of using the Fibonacci sequence to get approximations to τ. (*Hint:* Consider $\tau = 1 + \frac{1}{\tau}$ and Problem 3.)

*5 Show that $x = \frac{1}{\sqrt{5}}\left[\left(\frac{1 + \sqrt{5}}{2}\right)^n - \left(\frac{1 - \sqrt{5}}{2}\right)^n\right]$ where $n = 1, 2, 3, 4$, can be used to generate the first four Fibonacci numbers.

6 a) Use the quadratic formula to solve the equation $z^2 - 3z - 1 = 0$ for the positive root only. Express your answer to five decimal places. Use a calculator if possible.

b) Continued fractions and convergents can be used to approximate the positive root (provided one exists) of quadratic equations like $z^2 - 3z - 1$. This technique is illustrated below.

$$z^2 - 3z - 1 = 0$$

$$z - 3 - \frac{1}{z} = 0$$

$$z = 3 + \frac{1}{z}$$

Now, we form a continued fraction as follows:

$$z = 3 + \cfrac{1}{3 + \cfrac{1}{3 + \cfrac{1}{3 + \cfrac{1}{3 + \ldots}}}}$$

Form the first seven convergents $(3, 3 + \frac{1}{3}$, and so on) and their decimal equivalents correct to five decimals. Note how these values get closer and closer to your answer for part a.

7 Show how to generalize the procedure in part b of Problem 6 to get a continued fraction associated with any polynomial equation of the form $ax^2 + bx + c = 0$.

*8 a) Consider once again the characteristic equation $x^2 + x - 1 = 0$. Suppose we had attempted solution in the following way:

$$x^2 + x - 1 = 0$$

$$x^2 = 1 - x$$

$$x = \sqrt{1 - x} \quad \text{(considering the positive root only)}$$

Now replace x in the right member of this equation repeatedly by $\sqrt{1-x}$, thus producing the following infinite radical.

$$x = \sqrt{1 - \sqrt{1 - \sqrt{1 - \sqrt{1}}}} - \dots$$

Form the first six convergents

$$(\sqrt{1}, \sqrt{1 - \sqrt{1}}, \sqrt{1 - \sqrt{1 - \sqrt{1}}}, \text{ and so on})$$

Now simplify. What happens?

b) Now consider the equation $x^2 - x - 1 = 0$. In Section 4 you learned that the positive root of this equation has the value r. Follow the procedure given in part a to see if the first four successive convergents obtained from $x^2 - x - 1 = 0$ seem to be approaching the value of r.

**9 Write a computer program to generate the first 50 Fibonacci numbers.

10 Expand each of the following binomials.
a) $(a + b)^0$ c) $(a + b)^2$ e) $(a + b)^4$
b) $(a + b)^1$ d) $(a + b)^3$ f) $(a + b)^5$

11 To expand a binomial like $(a + b)^{12}$ would be quite a tedious task. However, there is a technique or formula that enables us to find more easily the expansion of $(a + b)^n$, where n is any whole number. To help us discover this technique, notice that coefficients of a, b, and ab terms

in each of the expansions in Problem 10 form the following pattern.

$$
\begin{array}{llllllll}
(a + b)^0 & 1 & & & & & \\
(a + b)^1 & 1 & 1 & & & & \\
(a + b)^2 & 1 & 2 & 1 & & & \\
(a + b)^3 & 1 & 3 & 3 & 1 & & \\
(a + b)^4 & 1 & 4 & 6 & 4 & 1 & \\
(a + b)^5 & 1 & 5 & 10 & 10 & 5 & 1
\end{array}
$$

This is the famous *Pascal triangle* — actually just a part of it (six rows to be exact), because it goes on forever. Notice that each row begins and ends with 1. Also, each row except the first and second rows can be determined from the preceding row. Take the fifth row, for example. The first and last numbers are 1, as in every other row. Observe that the second number in the fifth row (4) is the sum of the second number in the fourth row (3) and the first number in the fourth row (1). The third number in the fifth row (6) is the sum of the third number in the fourth row (3) and the second number in the fourth row (3). The fourth number in the fifth row (4) is the sum of the fourth number in the fourth row (1) and the third number in the fourth row (3). In other words, with the exception of the first and last numbers (both 1), each number in Pascal's triangle is the sum of the number directly above and the number above and directly to the left.

Suppose that we find the seventh row of Pascal's triangle; it is the following:

$$1 \quad 6 \quad 15 \quad 20 \quad 15 \quad 6 \quad 1$$

How do we use these numbers to write the expansion of $(a + b)^6$? You should have noticed in Problem 10 that the first term in each expansion is a raised to the power of the binomial (with the exception of $(a + b)^0$, which equals 1). In each successive term the exponent of a is reduced by 1, until in the last term $a^0 = 1$. Just the reverse pattern occurs for b. That is, the exponent of b in the first term is 0, in the second term it is 1, in the third term it is 2, and so on, until in the last term it is the number that is the power to which the binomial is being raised. Therefore, from the seventh row of Pascal's triangle we obtain the following expansion of $(a + b)^6$.

$$1a^6 b^0 + 6a^5 b^1 + 15a^4 b^2 + 20a^3 b^3 + 15a^2 b^4 + 6a^1 b^5 + 1a^0 b^6,$$

or

$$a^6 + 6a^5 b + 15a^4 b^2 + 20a^3 b^3 + 15a^2 b^4 + 6ab^5 + b^6$$

a) Expand Pascal's triangle through the thirteenth row. Then use this row to determine the expansion of $(a + b)^{12}$.

b) Examine the elements along the diagonals in the

Pascal triangle below. Then tell what relationship exists between Pascal's triangle and the Fibonacci numbers?

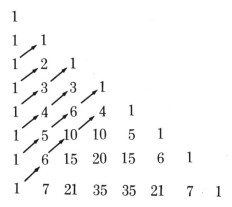

**12 Prove that it is impossible for any three different Fibonacci numbers to be the lengths of the sides of a triangle. (*Hint*: Consider two cases — either the Fibonacci numbers are consecutive or they are not consecutive.)

**13 Prove that if $\dfrac{a}{b} < \dfrac{c}{d}$, where a, b, c, and d are positive integers, then $\dfrac{a}{b} < \dfrac{a+c}{b+d} < \dfrac{c}{d}$.

**14 In the sequence $\dfrac{1}{1}, \dfrac{1}{2}, \dfrac{2}{3}, \dfrac{3}{5}, \dfrac{5}{8}, \dfrac{8}{13}, \dfrac{13}{21}$, ..., it appears as though each term is between the preceding two terms. Prove that this is always the case for this sequence.

PROJECTS ✳ ✳ ✳ ✳ ✳ ✳ ✳ ✳ ✳ ✳ ✳ ✳ ✳ ✳ ✳ ✳ ✳ ✳ ✳

Study Figure 14 very carefully and then try constructing a chart that will illustrate the family tree for the rabbits over the entire twelve-month period. You will have to arrange your chart on a large sheet of poster board or paper.

SECTION 9

A
Puzzling
Situation

In this section we will solve an interesting puzzle concerning a square and a rectangle. In so doing, we will study some relationships between Fibonacci numbers and squares and rectangles.

Referring to Figure 15 suppose that we were to cut pieces I, II, III, and IV from a square piece of cardboard whose dimensions are 8 × 8. We then rearrange the pieces to form the rectangle whose dimensions are 5 × 13. Does anything seem wrong? A moment's reflection reveals that there exists a discrepancy of one unit in the areas of the square and the rectangle. The square is one square unit smaller than the rectangle. But how can this be? After all, the same pieces make up both figures!

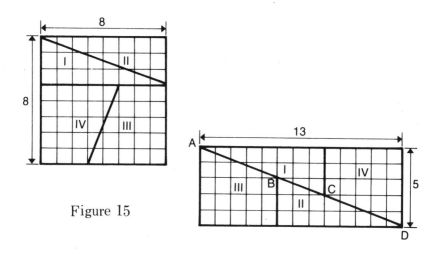

Figure 15

The answer to our puzzle is found in the fact that the points A, B, C, and D are not collinear, which we will prove in Problem 2. In fact, angles ABD and ACD are obtuse. Whether or not we will be able to observe that A, B, C, and D are not collinear or that ∢ABD and ∢ACD are obtuse will depend upon the units we use in making the cardboard models and the care with which we cut them out.

Now let us explore the dimensions of the square and the rectangle in more detail, that is the length of a side of the square and the length of the base and the height of the rectangle. Notice that these dimensions (5, 8, and 13) are *consecutive* Fibonacci numbers. Now consider Theorem 8, which concerns three consecutive Fibonacci numbers. The subscripts of F indicate the number of the term; for example; $F_1 = 1, F_2 = 1,$ $F_3 = 2, F_4 = 3, F_5 = 5,$ and so on. Then F_n would be the nth term of the Fibonacci sequence.

Theorem 8: $F_{n-1} \, F_{n+1} - F_n^2 = (-1)^n$, where $n > 1$.

This theorem states that for any three consecutive Fibonacci numbers, the difference between the product of the first and third numbers and the square of the second number is 1 if n is even and is -1 if n is odd.

Before proving this theorem, let us look at a few examples. The first eight Fibonacci numbers are given below.

F_1	F_2	F_3	F_4	F_5	F_6	F_7	F_8	$\cdots$
1	1	2	3	5	8	13	21	$\cdots$

Example 1: Suppose $n = 2$. Then $F_n = 1, F_{n-1} = 1, F_{n+1} = 2.$ By Theorem 8, $1 \cdot 2 - 1^2 = (-1)^2$, which we see is true.

Example 2: Let $n = 5$. Then $F_n = 5, F_{n-1} = 3, F_{n+1} = 8.$ By Theorem 8, $3 \cdot 8 - 5^2 = (-1)^5$, or $24 - 25 = (-1)^5$, which we see is true.

Example 3: Suppose $n = 4$. *Then* $F_n = 3, F_{n-1} = 2, F_{n+1} = 5.$ By Theorem 8, $2 \cdot 5 - (3)^2 = (-1)^4$, or $10 - 9 = (-1)^4 = 1$, which again is true.

We have just shown that the theorem holds for $n = 2$, 5 or 4. The theorem·says it holds for *all* positive integers greater than 1. How can we show that it does? This is an ideal situation for the method of proof known as *mathematical induction*.

Proof:

Let S be the set of all positive integers greater than 1. Let T be the set of all positive integers *that make the theorem true*. Our objective in this proof is to show that S = T. To show this using the principle of mathematical induction, we must establish two things:

(1) We must show that T contains 2, the first element in S. That is, we must show that the theorem is true when $n = 2$. We have already done this in Example 1.

(2) We must show that whenever T contains some element of S, say k, then it also contains the next element $k + 1$.

To establish (2) we will assume that k belongs to T; in other words, that $F_{k-1} F_{k+1} - F_k^2 = (-1)^k$ is true. Our task then becomes one of showing that when $n = k + 1$, then $F_k F_{k+2} - (F_{k+1})^2$ can be expressed as $(-1)^{k+1}$. This is established by the steps that follow.

$$F_k F_{k+2} - (F_{k+1})^2 = F_{k+2} F_k - (F_{k+1})^2$$
$$= (F_{k+1} + F_k)F_k - (F_{k+1})^2$$
$$= (F_k^2 + F_k F_{k+1}) - (F_{k+1})^2$$
$$= F_k^2 + [F_k F_{k+1} - (F_{k+1})^2]$$
$$= F_k^2 + F_{k+1}(F_k - F_{k+1})$$

But since $F_{k+1} = F_{k-1} + F_k$, it follows that $-F_{k-1} = F_k - F_{k+1}$.

Substituting, we have:

$$F_k F_{k+2} - (F_{k+1})^2 = F_k^2 + F_{k+1}(-F_{k-1}),$$
$$= (-1)(F_{k-1} F_{k+1} - F_k^2).$$

We know by inductive hypothesis that:

$$F_{k-1} F_{k+2} - F_k^2 = (-1)^k.$$

Therefore, upon substituting again, we get:

$$F_k F_{k+2} - (F_{k+1})^2 = (-1)(-1)^k$$
$$= (-1)^{k+1}.$$

Consequently: $F_k F_{k+2} - (F_{k+1})^2 = (-1)^{k+1}$.

This concludes the proof of part (2). Hence, by the principle of mathematical induction, T = S. That is, our theorem is true for *all* integers greater than 1.

This theorem can be interpreted as saying that given any three consecutive Fibonacci numbers, the difference between the area of a square whose side length is the middle Fibonacci number and the area of a rectangle whose dimensions are the other two Fibonacci numbers will always be one square unit. A close inspection of the theorem reveals that whenever the square has a side length that is a Fibonacci number with an even subscript, then the area of the rectangle will always exceed the area of the square by one square unit. What will be the case when the square has a side length that is a Fibonacci number with an odd subscript (See Problem 1)?

Although it does not show up in Figure 15, it has already been suggested in the explanation of the flaw in the puzzle that the extra unit is gained (or lost) along the diagonal of the rectangle. A distorted drawing of the way things really are, when the extra unit is gained, is given below where, for the rectangle in Figure 15, $F_{2n} = F_6 = 8$, $F_{2n-1} = F_5 = 5$ and $F_{2n+1} = F_7 = 13$.

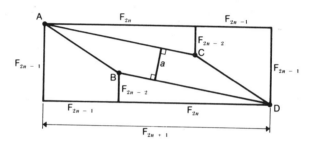

Figure 16

PROBLEMS ??

1 Verify for several examples that when the middle of three consecutive Fibonacci numbers has an odd subscript, the area of the rectangle will be one square unit less than the area of the square.

2 Use trigonometric functions to show that in the rectangle in Figure 15 points A, B, C, and D are not collinear. (*Hint*: Show that $m \angle ABD \neq 180$.)

3 Show that for larger and larger Fibonacci numbers, the parallelogram (See Figure 16) will become less and less noticeable. Do this by establishing that the altitude a of the parallelogram is related to the Fibonacci numbers by

$$a = \frac{1}{\sqrt{F_{2n}^2 + (F_{2n-2})^2}}.$$

4 The puzzle discussed in this section need not have any relationship to the Fibonacci numbers. To see what is meant by this statement, suppose we select any two real numbers, restricting ourselves to positive real numbers in order to make the side lengths of the square and rectangle seem reasonable. Let us use 4 and 6. Now we generate a portion of the *additive sequence*, say 4, 6, 10, 16, 26, 42, and 68. Select all consecutive triples of numbers from this portion of the progression and compute the product of the first and the third numbers minus the square of the second number. Is the absolute value of the difference obtained in each case the same?

5 Interpret the results in Problem 4 geometrically.

6 Letting $N_1 = 4$, $N_2 = 6$, $N_3 = 10$, $N_4 = 16$, $N_5 = 26$, $N_6 = 42$, and $N_7 = 68$, construct two cardboard squares like the one in Figure 15. Let one square have the dimensions of N_i, where i is an even subscript. Let the other square have dimensions of N_i, where i is an odd subscript. Now rearrange the pieces from the two squares to form two rectangles. Is it obvious from your models where the units (in this case 4) are lost or gained?

7 Generate a few terms of several additive sequences by starting with any two real numbers that you wish. Follow the verification procedure described in Problem 4.

Problems 4 through 7 suggest that all additive sequences can form the basis for a puzzle like the one described. However, there is *one* exception to this statement. That is, there is one additive sequence such that if we select any three consecutive terms of the sequence, multiply the first and third terms and subtract the square of the middle term, we will get zero.

Geometrically, of course, this would mean that the square and rectangle whose dimensions were these consecutive numbers would have exactly the same area. This sequence is generated by starting with 1 and τ.

$$1, \tau, 1 + \tau, 1 + 2\tau, 2 + 3\tau, 3 + 5\tau, 5 + 8\tau, \ldots$$

The calculations below will enable us to put this sequence into what turns out to be a more useful form.

$\tau^2 = 1 + \tau$ (This follows from Problem 5, Section 4.)

$\tau^3 = \tau^2\tau = (\tau + 1)\tau = \tau^2 + \tau = \tau + 1 + \tau = 1 + 2\tau$

$\tau^4 = \tau^3\tau = (2\tau + 1)\tau = 2\tau^2 + \tau = 2\tau + 2 + \tau = 2 + 3\tau$

$\tau^5 = \tau^4\tau = (3\tau + 2)\tau = 3\tau^2 + 2\tau = 3\tau + 3 + 2\tau = 3 + 5\tau$

$\tau^6 = \tau^5\tau = (3 + 5\tau)\tau = 3\tau + 5\tau^2 = 3\tau + 5(\tau + 1) = 3\tau + 5\tau + 5 = 5 + 8\tau$

$\vdots$

Notice that the third term of the additive sequence, $1 + \tau$, is equal to τ^2; the fourth term, $1 + 2\tau$, is equal to τ^3; the fifth term $2 + 3\tau$, is equal to τ^4; and so on. Therefore, by substituting in

$$1, \tau, 1 + \tau, 1 + 2\tau, 2 + 3\tau, 3 + 5\tau, 5 + 8\tau, \ldots,$$

we obtain the sequence: $\quad 1, \tau, \tau^2, \tau^3, \tau^4, \tau^5, \tau^6, \ldots$

It is interesting to note that in this new sequence any three consecutive terms will have the form t_{n-1}, t_n, t_{n+1} where $n \geq 2$,

and where $t_{n-1} = \tau^{n-2}$, $t_n = \tau^{n-1}$, and $t_{n+1} = \tau^n$. It is clear then that

$$t_{n-1} t_{n+1} - t_n^2 = \tau^{n-2}\tau^n - \tau^{2n-2} = 0.$$

SECTION 10

Dynamic Symmetry

The field of art may seem like a strange place for mathematics to appear. Nevertheless, mathematics has played an important role in art, particularly the art of the early Greeks and Egyptians. Socrates (who was a professional sculptor as well as a philosopher) describes the ancient Greek point of view concerning the relationship between mathematics and art when he says:

> "If arithmetic, mensuration and weighing be taken away from
> any art, that which remains will not be much—the rest will be
> only conjecture, and the better use of the senses which is
> given by experience and practice, in addition to a certain power
> of *guessing,* which is *commonly called art.*" (Bowes, 12)

The use of mathematics in the art and architecture of the early Greeks and Egyptians refers to much more than just linear measurements. It involves a subtle concept known as *dynamic symmetry*. The purpose of this section is to introduce us to this concept.

To begin our discussion, let us distinguish between *static* and *dynamic symmetry*. The following succinct quotes serve to point out this distinction:

"Since the days of classic art, artists have used what is called static symmetry. The dimensions of its designs are commensurable. Their lengths and widths have definite arithmetic measures. The geometry of static design consists of regular geometric figures, perhaps superposed at different angles, as two squares put one on the other at an angle of 45°. The art of the classic periods in Egypt and Greece used a more subtle proportion, as was recently discovered by the late Jay Hambidge of Yale University. He gave it the name of dynamic symmetry. In designs of this proportion, the dimensions are incommensurable. The most common ratios are $1: \sqrt{2}$, and $1: \sqrt{3}$, and $1: \sqrt{5}$. Dynamic designs are evolved by the use of areas rather than by line measurements. They are based on root rectangles which have the foregoing ratios." (Gugle, 26)

". . . Symmetry for the artist means the arrangement of elements around an axis which passes through the center of the composition. The symmetric effect thus produced may be static or it may be dynamic. To achieve *static* symmetry, one employs designs of definite, fixed arithmetic lengths.

Thus, concentric circles, where the radii constantly increase by fixed amounts, are symmetric in a static sense. As the name implies an effect of stillness and quiet is produced.

Dynamic symmetry in opposition to static symmetry makes use of proportions which cannot be measured in length i.e. their dimensions are incommensurable, but which are commensurable in areas . . . . Dynamic means growth, power, and movement. It gives animation and life to an *artist's* work. This movement evolves from the designs of dynamic symmetry which employ areas rather than lines." (Law, 39)

These quotes indicate that it is the proportioning of areas, rather than lines that distinguishes dynamic symmetry from static symmetry.

In the quotations reference is made to what are called root rectangles. Actually, we have already encountered root rectangles; for example, in Problem 4, of Section 7. Now we will construct one.

In Figure 17 let ABCD be a unit square. Draw the diagonal $\overline{AC}$ and mark off its length on $\overline{AB}$ extended; then $AE = \sqrt{2}$. If we now construct the rectangle having $\overline{AE}$ and $\overline{AD}$ as sides, we have what is called a root-two rectangle.

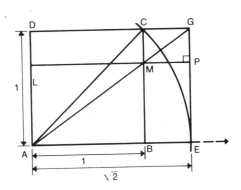

Figure 17

A feature of the root-two rectangle is that the area of the square constructed on a longer side is an integral multiple of the area of the square constructed on one of the short sides. That is, the areas of the two squares are commensurable. In fact, this is a property of all root rectangles and it is this commensurability of areas which lies at the heart of the artist's and architect's use of dynamic symmetry.

Referring to Figure 17 again, consider diagonal $\overline{AG}$. Call M the point of intersection of $\overline{AG}$ and $\overline{CB}$. Through M construct a line perpendicular to $\overline{GE}$. We observe that in the root-two rectangle the length of $\overline{AL}$ is approximately equal to 0.7071 (See Problem 7.), while LP equals approximately 1.4142, or twice the length of $\overline{AL}$. However, the sides of the root-two rectangle ($\overline{AD}$ and $\overline{AE}$), have lengths 1 and $\sqrt{2}$. That is, the lengths of the sides are in the ratio $1:\sqrt{2}$. These mathematical facts can be of great value in design. The artist who is skilled in the use of dynamic symmetry can take advantage of such numerical relationships in creating designs for paintings and other layouts.

Now let us construct a root-five rectangle. We begin by constructing a unit square ABCD.(See Figure 18.)

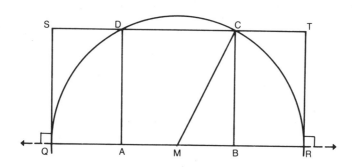

Figure 18

Locate the midpoint of $\overline{AB}$ and call it M. Draw $\overline{MC}$. Using M as the center and MC as a radius, draw a semicircle intersecting $\overline{AB}$ extended at points Q and R. At Q and R, complete the rectangle. SQRT is a root-five rectangle; that is, its side lengths are in the ratio $1:\sqrt{5}$ and hence are incommensurable. Squares constructed on the sides of this rectangle have areas whose ratio is 1:5. Notice that these areas are commensurable.

An observation follows from the construction of this rectangle. Look again at Figure 18. In constructing the root-five rectangle SQRT, these two rectangles were formed: SQBC and DART. What kind of rectangles are they? (See Problem 6.)

The root-five rectangle serves as the basis for a high type of design. Figure 19 shows a Greek drinking cup that is in the Boston Museum of Fine Arts. BGFC is a root-five rectangle, which means that BGKL and MJFC are whirling square rectangles, that is, Golden Rectangles. The width of the foot $\overline{JK}$ is the length of a side of the Golden Rectangles. Note also that diagonals $\overline{JC}$ and $\overline{BK}$ of the Golden Rectangles meet at a point that marks the division of the bowl and the foot of the cup. For further analysis of the dynamic symmetry of this cup you may wish to consult the references of Bowes (12) and Hambidge (28).

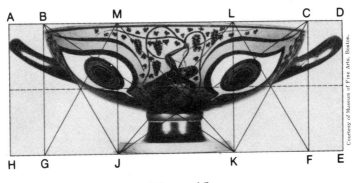

Figure 19

Further illustrations of the use of dynamic symmetry in art can be found in the paintings of some "modern-day" artists. Some good examples are found in the work of Seurat (1859–1891). Two of his paintings are shown here. Along with the paintings are diagrams which show a partial analysis of the works. Notice in the analyses how Seurat made use of the Golden Section in laying out his canvases. The numbers τ and $\tau\sqrt{\tau}$ written inside the rectangles in the first analysis represent the ratio of the longer to the shorter side of the various rectangles. Matila Ghyka in her book *The Geometry of Art and Life* (24) claims that the hypnotic effect of Seurat's paintings is due to this rigorous mathematical technique of composition.

Seurat: *The Circus*, 1891

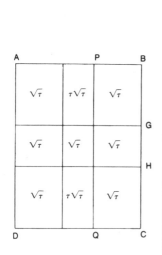

Figure 20

Seurat: *Parade*, 1889

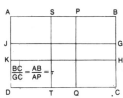

Figure 21

It is hoped that by this point you are getting some idea of what dynamic symmetry is and how some of the ideas introduced earlier in this book fit into this concept. We will close this section by considering the words of the late Jay Hambidge who rediscovered the use of dynamic symmetry in Greek art:

"In the beginning, the Greeks used Static Symmetry, but some time between the 6th and 7th Centuries B.C., analysis of vases shows that knowledge of Dynamic Symmetry was secured from the Egyptians, and it gradually supplanted the older scheme, until by the beginning of the 5th Century B.C., about eighty-five percent of all artists were using it exclusively, about twelve percent apparently were never initiated and used the old schemes, and about three percent had no design knowledge whatever . . . the Greek sculptor awoke to the fact that this wonderful design principle could be applied to the human figure. Immediately Greek sculpture was changed and it started on the movement that reached the full and glorious fruit of the Golden Age of the World's art; this age of Pericles." (Bowes, 12)

PROBLEMS ???

1 Construct the following:
   a) a root-three rectangle
   b) a root-four rectangle

2 What is the feature that makes all root rectangles important from an artist's standpoint?

*3 Consider the root-two rectangle in Figure 17. We know from geometry that $\overline{AG}$ divides rectangle AEGD into two congruent triangles of equal area.
   a) What does $\overline{AG}$ do to the area of ABML?
   b) What does $\overline{AG}$ do the the area of CMPG?
   c) How must area DLMC and MPEB be related?
   d) Prove your answer to part c.
   e) How are the areas of ABCD and AEPL related?

4 Study again the construction procedure for Figure 18 and then show that the length of $\overline{QR}$(and thus $\overline{ST}$) is $\sqrt{5}$.

5 Verify that in Figure 18 a square constructed on $\overline{QR}$ has an area five times that of a square constructed on $\overline{BC}$.

**6 In Figure 18 draw in $\overline{QC}$ and $\overline{CR}$. In doing this, it follows that $\triangle QCR \sim \triangle QBC \sim \triangle CBR$. It also follows that $\dfrac{QB}{BC} = \dfrac{BC}{BR}$. Use this information to prove that B produces the Golden Section on $\overline{AR}$. (*Hint:* Do not assume AB = 1.)

*7 Prove that the length of $\overline{LA}$ in Figure 17 is approximately 0.7071.

SECTION 11

The
Golden Section
In Nature

In this final section we will see how some of the topics we have studied find their way into nature. Probably most of us have never taken the time to examine very carefully a twig of a tree, a sunflower, a pine cone, or the number or arrangement of petals on a flower. If we were to do this, several things would become apparent.

First, we would find that the number of petals on a flower blossom is often one of the Fibonacci numbers. See Hoggatt (34). It is also interesting to note that some leaves, flowers, and animals have shapes that can be circumscribed by the regular pentagon. Some examples are shown in Figure 22. Also Walt · Disney's film *Donald in Mathmagic Land* (51) has some excellent illustrations of the pentagon in nature.

Figure 22

Secondly, if we were to examine the way in which leaves or buds are arranged around a twig of a tree (In botany the study of leaf arrangements on a twig is known as *phyllotaxis.*), we would notice that in most cases the ratio of the number of turns to the number of leaves is a Fibonacci number. For example, suppose that we were to select a leaf at random and assign it the number 0 as shown in Figure 23. Then we start moving around the stem clockwise or counterclockwise until we encounter the first leaf that is directly above leaf 0. We then divide the number of leaves we have encountered on our trip (including the final one) into the number of complete revolutions of the stem that we made. In Figure 23 the leaf arrangement can be represented by the fraction $\frac{1}{3}$.

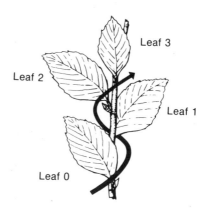

Figure 23

The elm tree has the $\frac{1}{2}$ arrangement and the beech tree has the $\frac{1}{3}$ arrangement. Some trees have the $\frac{3}{8}$ arrangement; certain bushes have the $\frac{5}{13}$ arrangement. As we have mentioned, these numbers will often be Fibonacci numbers as numerators and denominators.

Another striking example of the occurrence of Fibonacci numbers in nature is offered by the spiral pattern of the seeds of a sunflower and the scale patterns of a pine cone. If we were to look at the head of a sunflower, we would see two sets of spirals. One set of spirals radiates from the center in a clockwise direction, while the other radiates in a counterclockwise direction. The usual sunflower head (5 to 6 inches in diameter) contains 34 spirals going in one direction and 55 going in the other direction. For smaller or larger sunflowers this ratio changes, but almost always remains the ratio of two Fibonacci numbers.

In Figure 24 we can see the thirteen clockwise spirals of the pine cone along with the counterclockwise spirals.

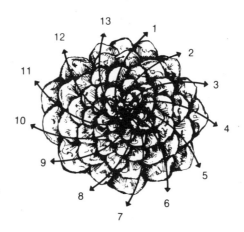

Figure 24

Also, fresh pineapples usually have their number of whorls equal to Fibonacci numbers. For an analysis of these whorls, see Coxeter (18).

It has been verified through many observations and measurements that not only are the numbers of these spirals usually Fibonacci numbers, but, in addition, they form the whirling squares spiral mentioned in Section 7. This same spiral also can be seen by examining many different animal horns. See Thompson (48).

The nautilus shown in Figure 25 furnishes another good example of the logarithmic spiral's occurence in nature.

Figure 25

As a final and more intimate illustration of the Golden Section in nature, consider the ratio of a person's total height to his navel height (height from navel to bottom of foot). Some people have actually conducted experiments in which these measurements were made on a number of people. See Gardner (22). You guessed it! The ratio turned out to be very close to 1.618!

PROBLEMS ???

1 For a number of different trees growing in your area
determine the ratio of the number of turns to the number of
leaves. Did you obtain ratios of Fibonacci numbers?

2 Find as many different examples as you can of the pentagon
in nature.

3 Examine various kinds and sizes of pine cones and sunflowers. Carefully count the number of clockwise and
counterclockwise spirals. Do these numbers turn out to be
Fibonacci numbers?

4 Determine the ratio of total height to navel height of some
of your classmates. Do the ratios more nearly equal
1.618 for the boys or the girls?

APPENDIX

APPENDIX A

AN ALTERNATE WAY OF CONSTRUCTING THE GOLDEN SECTION

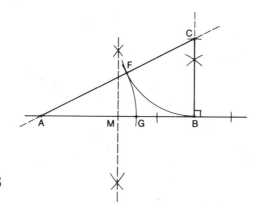

Figure 26

PROCEDURE:

Draw a line segment $\overline{AB}$. Locate the midpoint of $\overline{AB}$ and call it M. At point B, construct $\overline{BC}$ perpendicular to $\overline{AB}$ and such that BC = MB. Draw $\overline{AC}$ and locate F on $\overline{AC}$ such that CB = CF. Then locate G on $\overline{AB}$ such that AF = AG. G is the required point of division.

PROOF:

Let $\overline{AB}$ have an arbitrary length of $2a$. Then CB = a. Hence AC = $\sqrt{5a^2}$ = $a\sqrt{5}$. Since FC = CB = a, it follows that AF = $a\sqrt{5} - a$ = AG. So GB = $2a - (a\sqrt{5} - a)$ = $3a - a\sqrt{5}$. To show that AG^2 = AB · GB we make the following replacements: AG = $a\sqrt{5} - a$, AB = $2a$, and GB = $3a - a\sqrt{5}$. Since $(a\sqrt{5} - a)^2$ = $(2a)(3a - a\sqrt{5})$, our definition guarantees us that G is the desired point.

APPENDIX

APPENDIX B

A MASCHERONI CONSTRUCTION OF THE GOLDEN SECTION

Traditionally, all construction problems in geometry are done with the aid of an unmarked straightedge and a compass. Such restrictions, as mentioned in Section 3, are credited to Plato. In 1797, however, the Italian mathematician Lorenzo Mascheroni published a book entitled *Geometria del Compasso*. Mascheroni showed in this book that all constructions that could be done with a straightedge and compass also could be done with just a compass. Below is a so-called Mascheroni construction that shows how to divide $\overline{AB}$ into extreme and mean ratio using only a compass.

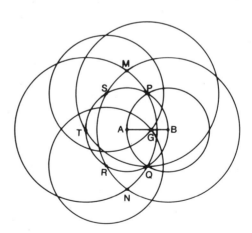

Figure 27

PROCEDURE:

Construct two circles having A and B as centers and AB as radius. Call their points of intersection P and Q. Use point P as center and with radius PQ draw a third circle that intersects the circle with center A; call the point of intersection T. Now use B as center and PQ again as radius to construct a fourth circle that intersects the circle with center A; call the points of intersection R and S. Next use T as center and PQ as radius to draw circles intersecting the circle with center B and radius PQ at points M and N. Then using AM as radius and S and R as centers, construct two circles meeting at G. G will be on $\overline{AB}$ and it is our desired point.

PROOF:

To prove that the Mascheroni construction really does produce a Golden Section we need to show that $\dfrac{AB}{AG} = \dfrac{AG}{GB}$. In order to show that $\dfrac{AB}{AG} = \dfrac{AG}{GB}$ we extend our construction by drawing right triangle SGV with another construction. See Figure 28. (In order to make the proof easier to follow, we have omitted all circles, except two, from the drawing in Figure 27.)

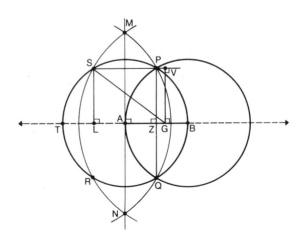

Figure 28

APPENDIX

We will now express the lengths of $\overline{AG}$ and $\overline{GB}$ in terms of AB. In $\triangle ZBP$, $PZ^2 = \dfrac{3AB^2}{4}$. In $\triangle ABM$ $MA^2 = MB^2 - AB^2$. But

MB = PQ, and PQ in terms of AB is $2 \cdot PZ = 2\sqrt{\dfrac{3AB^2}{4}}$. So

$MA^2 = 3AB^2 - AB^2 = 2AB^2$. By construction, MA = SG, so $SG^2 = MA^2 = 2AB^2$. In $\triangle SVG$, VG = PZ, so $VG^2 = PZ^2 = \dfrac{3AB^2}{4}$.

Hence, by the Pythagorean theorem, $SV = \sqrt{SG^2 - VG^2}$, or

$\sqrt{2AB^2 - \left(\dfrac{3AB^2}{4}\right)} = \sqrt{\dfrac{5AB^2}{4}} = \dfrac{AB}{2}\sqrt{5}$. Now SV = LG. Also,

$AL = AZ = \dfrac{AB}{2}$. So $AG = LG - AL = \left(\dfrac{AB}{2}\right)\sqrt{5} - \dfrac{AB}{2} = \left(\dfrac{AB}{2}\right)(\sqrt{5} - 1)$.

Hence GB = AB − AG or $AB - \left(\dfrac{AB}{2}\right)(\sqrt{5} - 1) = \left(\dfrac{AB}{2}\right)(3 - \sqrt{5})$.

Therefore, $\dfrac{AB}{AG}$ and $\dfrac{AG}{GB}$ become

$$\dfrac{AB}{\dfrac{AB(\sqrt{5} - 1)}{2}} \qquad \text{and} \qquad \dfrac{\dfrac{AB(\sqrt{5} - 1)}{2}}{\dfrac{AB(3 - \sqrt{5})}{2}}, \text{ respectively.}$$

A bit of manipulation shows the first fraction equal to

$\dfrac{2}{\sqrt{5} - 1} = \dfrac{-1 - \sqrt{5}}{-2} = \dfrac{1 + \sqrt{5}}{2} = \tau$, and the second fraction

equal to $\dfrac{2 + 2\sqrt{5}}{4} = \dfrac{1 + \sqrt{5}}{2} = \tau$. Hence, $\dfrac{AB}{AG} = \dfrac{AG}{AB}$. So, by

definition, G divides $\overline{AB}$ into a Golden Section.

APPENDIX

APPENDIX C

PROOF OF THE VALIDITY OF THE CONSTRUCTION OF THE REGULAR PENTAGON

In order to prove that the construction procedure used in Figure 5 really does produce a regular pentagon, we will first obtain a value for the cosine of θ, where θ is an angle such that $5\theta = 360$. After we have determined this value for $\cos \theta$, we will show that the construction in Figure 5 produces an angle PKG whose cosine is this same value.

Consider the inscribed pentagon in Figure 29.

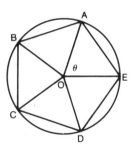

Figure 29

Assume that ABCDE is a regular pentagon. Then $m \angle AOE$ (denoted by θ) is such that $5\theta = 360$. From trigonometry we know that $\cos 5\theta = \cos 360 = 1$. Furthermore, $\cos 5\theta$ can be written as $\cos (2\theta + 3\theta)$, which by a well-known trigonometric identity equals $\cos 2\theta \cos 3\theta - \sin 2\theta \sin 3\theta$. That is, $\cos 5\theta = \cos (2\theta + 3\theta) = \cos 2\theta \cos 3\theta - \sin 2\theta \sin 3\theta$. Now we will rewrite the right side of this equation in terms of sines and cosines of θ. That means we will need equivalent expressions for $\cos 2\theta$, $\sin 2\theta$, $\cos 3\theta$, and $\sin 3\theta$ in terms of sines and cosines of θ. These equivalent expressions are given below.

$$\cos 2\theta = \cos^2 \theta - \sin^2 \theta$$

so, $\boxed{\cos 2\theta = 2 \cos^2 \theta - 1}$

or $\boxed{\cos 2\theta = 1 - 2 \sin^2 \theta}$

$\boxed{\sin 2\theta = 2 \sin \theta \cos \theta}$

$$\begin{aligned}
\cos 3\theta &= \cos 2\theta \cos \theta - \sin 2\theta \sin \theta \\
&= (2 \cos^2 \theta - 1) \cos \theta - 2 \sin^2 \theta \cos \theta \\
&= (2 \cos^2 \theta - 1) \cos \theta - 2(1 - \cos^2 \theta) \cos \theta
\end{aligned}$$

so, $\boxed{\cos 3\theta = 4 \cos^3 \theta - 3 \cos \theta}$

$$\begin{aligned}
\sin 3\theta &= \sin 2\theta \cos \theta + \cos 2\theta \sin \theta \\
&= 2 \sin \theta \cos^2 \theta + (1 - 2 \sin^2 \theta) \sin \theta \\
&= 2 \sin \theta (1 - \sin^2 \theta) + (1 - 2 \sin^2 \theta) \sin \theta
\end{aligned}$$

so, $\boxed{\sin 3\theta = 3 \sin \theta - 4 \sin^3 \theta}$

By substituting in the right side of $\cos 5\theta = \cos 2\theta \cos 3\theta - \sin 2\theta \sin 3\theta$, we get the following:

$$\begin{aligned}
\cos 5\theta &= (2 \cos^2 \theta - 1) (4 \cos^3 \theta - 3 \cos \theta) \\
&\quad -(2 \sin \theta \cos \theta)(3 \sin \theta - 4 \sin^3 \theta) \\
&= (2 \cos^2 \theta - 1) (4 \cos^3 \theta - 3 \cos \theta) \\
&\quad -(2 \sin^2 \theta \cos \theta)(3 - 4 \sin^2 \theta) \\
&= (2 \cos^2 \theta - 1) (4 \cos^3 \theta - 3 \cos \theta) \\
&\quad - [2(1 - \cos^2 \theta) \cos \theta](-1 + 4 \cos^2 \theta) \\
&= 16 \cos^5 \theta - 20 \cos^3 \theta + 5 \cos \theta.
\end{aligned}$$

APPENDIX

Recall that $\cos 5\theta = 1$ or $\cos 5\theta - 1 = 0$. Hence,

$$16 \cos^5\theta - 20 \cos^3\theta + 5 \cos \theta - 1 = 0.$$

If we let $\cos \theta = x$, then we have

$$16 x^5 - 20 x^3 + 5 x - 1 = 0,$$

which is equivalent to

$$(x - 1) (16x^4 + 16x^3 - 4x^2 - 4x + 1) = 0.$$

Therefore,

$$x - 1 = 0 \text{ or } 16x^4 + 16x^3 - 4x^2 - 4x + 1 = 0.$$

Suppose that $x - 1 = 0$, then $x = 1$.

Suppose that

$$16x^4 + 16x^3 - 4x^2 - 4x + 1 = 0,$$

then

$$(4x^2 + 2x - 1)^2 = 0, \ 4x^2 + 2x - 1 = 0 \text{ and } x = \frac{-1 \pm \sqrt{5}}{4}.$$

However, in Figure 29 angle θ is acute, so $x = \dfrac{-1 + \sqrt{5}}{4}$

rather than $\dfrac{-1 - \sqrt{5}}{4}$.

Hence, we have $\cos \theta = 1$ and $\cos \theta = \dfrac{-1 + \sqrt{5}}{4}$.

We will disregard our first solution (that is, $\cos \theta = 1$) because this means that θ must be some integral multiple of 360, which is impossible since $5\theta = 360$. We find, therefore, that in a regular pentagon, the cosine of one of the central angles can be expressed as $\cos \theta = \dfrac{\sqrt{5} - 1}{4}$.

Now that we have determined a value for cos θ, where θ is a central angle of the inscribed regular pentagon in Figure 29, all that remains is for us to show that the construction procedure illustrated in Figure 4 produces an angle PKG such that cos PKG $= \dfrac{-1 + \sqrt{5}}{4}$. Having done this, we will have validated the procedure for producing a regular pentagon.

In Figure 5 let KG $= 1$. Then KM $= \dfrac{1}{2}$. Hence MG $= \dfrac{\sqrt{5}}{2}$. From geometry we know that bisector $\overline{\text{MQ}}$ of $\sphericalangle$KMG divides

the side $\overline{\text{KG}}$ into segments $\overline{\text{KQ}}$ and $\overline{\text{QG}}$ such that $\dfrac{\text{KQ}}{\text{QG}} = \dfrac{\text{KM}}{\text{MG}}$.

It follows that $\dfrac{\text{KQ}}{1 - \text{KQ}} = \dfrac{\dfrac{1}{2}}{\sqrt{\dfrac{5}{2}}}$. Solving for KQ, we find that

KQ $= \dfrac{\sqrt{5} - 1}{4}$. Since $\triangle$KQP is a right triangle,

$$\cos \text{PKG} = \frac{\text{KQ}}{\text{KP}} = \frac{\left[\dfrac{\sqrt{5} - 1}{4} \right]}{1} = \frac{\sqrt{5} - 1}{4} = \frac{-1 + \sqrt{5}}{4}.$$

Comparing this value to the one above, we see that the procedure for constructing a regular pentagon shown in Figure 5 is valid.

It is interesting to note that the cosine of a central angle of a regular pentagon turns out to equal $\dfrac{1}{2\tau}$.

APPENDIX

APPENDIX D

ADDITIONAL PROBLEMS

NOTE: For answers to problems, refer to *Answers* section, p. 159.

PROBLEM 1

How many different triangles does the figure below contain? Congruent triangles are to be considered different if they occupy a different position in the figure. For example, PFA and AGT are to be counted as two different triangles even though they are congruent.

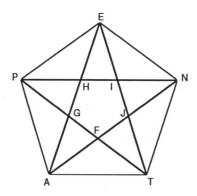

Figure 30

APPENDIX

PROBLEM 2

In Figure 31 below, let AC = 1. Figure 31 is a duplication of Figure 7a, page 22. Show how the numerical value of AB/AC can be derived more directly than was done on page 22.

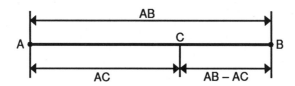

Figure 31

APPENDIX

PROBLEM 3

Prove that the following procedure describes a way of constructing (using only a compass and straightedge) a regular pentagon whose side is any given length. (A scientific calculator will be useful in this problem.)

Suppose, for example, that we wish to construct a regular pentagon having a side whose length is AB shown below:

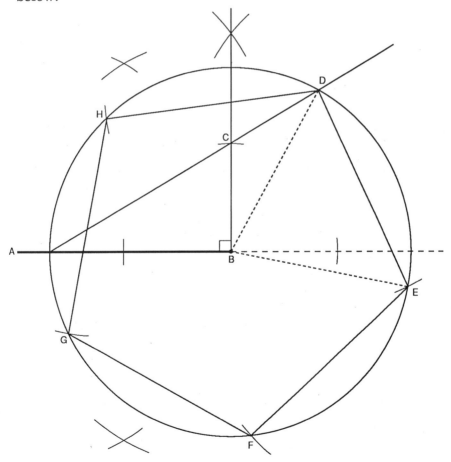

Figure 32

CONSTRUCTION PROCEDURE:

At point B, construct $\overline{BC}$ perpendicular to $\overline{AB}$ and length equal to AB/2. Draw $\overline{AC}$ and extend it to D such that BC = CD. Using B as the center of a circle and BD as the length of a radius, construct a circle. Use a compass to measure the length of $\overline{AB}$. This distance can be used to mark off exactly five points on the circumference of the circle starting (for example) at point D. Connecting the points D, E, F, G, and H in order produces a regular pentagon having a side length equal to AB.

PROBLEM 4
HOW TO FOLD A FIVE-POINTED STAR

1. Begin with a full sheet of paper, ABCD, as shown on page 110.

2. Fold it in half along the dotted line EF, and crease sharply.

3. Locate Z the midpoint of EF. Make a crease along ZG as shown such that angle EZF is one-half the size of angle FZG (36°). (A protractor is very helpful in determining the position of ZF and hence the crease ZG.)

4. Now fold edge ZG along edge ZF and crease sharply along edge ZH.

5. Fold ZE onto ZH and crease sharply.

6. Locate points M and N at equidistances from Z along creases ZG and ZH. Drop a perpendicular MP to Z.

    Locate point Q anywhere between P and Z. Cut along MQ and unfold to create a five-pointed star. Fixing point Q appropriately will produce a "perfect" star, or pentagram. (NOTE: If you cut along MN and unfold, you will get a regular decagon. If you cut along MP and unfold, you will get a regular pentagon.)

    Experiment using the above procedures to generate a variety of decagons, pentagons, and stars. How do you suppose one could precisely locate point Q so that a "perfect" star (a pentagram) would be produced?

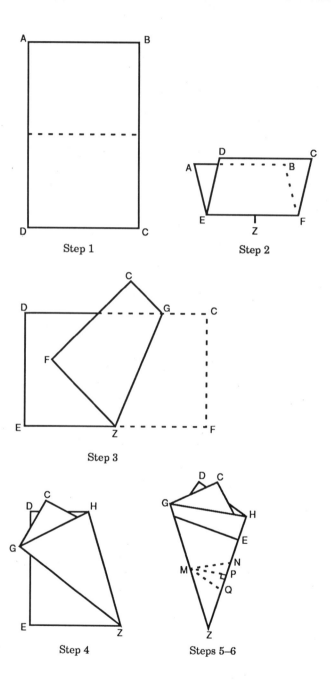

Step 1

Step 2

Step 3

Step 4

Steps 5–6

APPENDIX

PROBLEM 5

Begin this problem by placing *any* two squares next to one another, such as ABQP and BCLM. Squares whose side lengths are 1 and 2 have been arbitrarily chosen. Next, complete the rectangle ACLN by extending ML and AP until they intersect at N. Calculate the ratio of AC/CL. Now form the larger rectangle ACJI. Create its long side by extending lines CL and AN AC units in length. Calculate the ratio CJ/AC. Form the rectangle ADKI similarly and calculate the value of the ratio AD/DK. Continue forming larger and larger rectangles. What do you observe about the numerical value of the ratios of the longer to the shorter sides of these successively larger rectangles?

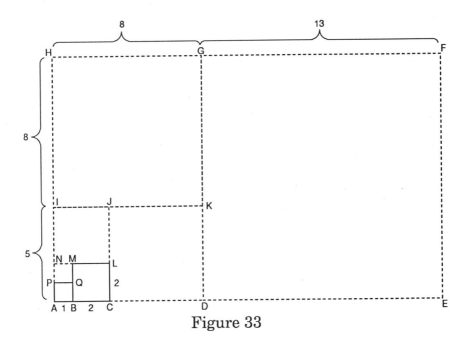

Figure 33

APPENDIX

PROBLEM 6

The last project of Section 7 (page 59) illustrates a pattern from which an icosahedron may be folded. The icosahedron forms part of the logo (shown below) of the Mathematical Association of America.

A tongue-in-cheek article by Branko Grünbaum in the January 1985 issue of the MAA's *Mathematics* Magazine describes a clever way of constructing a diagram of an icosahedron. See if you can follow the simple directions below to generate an icosahedron similar to the one shown.

Directions: Construct a regular hexagon ABCDEF (as shown in Figure 34 below), and locate its center O. Then construct the six points (such as G) on the six radial segments $\overline{OF}$, $\overline{OE}$, $\overline{OD}$, $\overline{OC}$, $\overline{OB}$, and $\overline{OA}$ that form the Golden Section on the segments. That is, OG/GF = τ. Connect the twelve points (A, B, C, D, E, F, G, H, I, J, K, and L) as shown to form a two-dimensional diagram of an icosahedron.

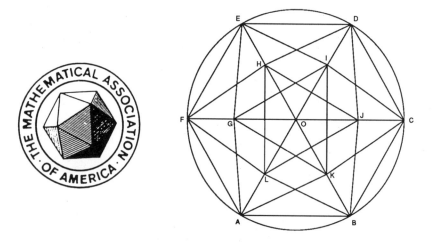

Figure 34

APPENDIX

PROBLEM 7

Shown below is a pattern for making what Seitz refers to as a golden cuboid. It is defined as a rectangular solid (namely a rectangular *parallelepiped*, or prism whose bases are parallelograms) whose length, width, and height are in the ratio $\tau : 1 : (1/\tau)$. Prove that the ratio of the surface areas of any golden cuboid to that of a sphere that circumscribes it is $\tau : \pi$.

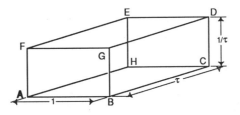

Figure 35

APPENDIX

PROBLEM 8

The following problem is an interesting application of the Fibonacci Sequence and the Golden Section: The maintenance department of a large public aquarium has cleaned its tanks and is now refilling them. First an integral amount of water is pumped in, x gallons. Then a second integral amount of water is pumped in, y gallons. The additional amounts of water added to the aquarium will always be the sum of the previous two amounts. The aquarium is full after the twentieth addition of water— exactly 1,000,000 gallons. How many gallons were in the first two amounts pumped in, x and y?

PROBLEM 9

Much has been written in this book regarding the Golden Section's relationship to regular pentagons and pentagrams. The non-regular pentagram shown below has many of its angles numbered. Find the sum of the measures of angles 1, 2, 3, 4, and 5. Does the answer to this problem change if a different pentagram is used? NOTE: This problem can be solved without using any knowledge of the Golden Section.

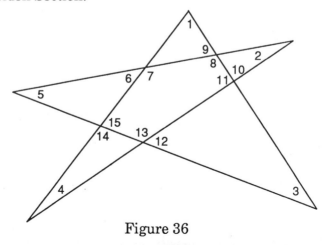

Figure 36

Answers

SECTION 1 (pages 1–4)

1   **a)** 4   **b)** 6   **c)** $3\sqrt{5}$ **d)** 9

2   **a)** $2\sqrt{6}$ **b)** $4\sqrt{2}$ **c)** $\sqrt{41}$ **d)** $5\sqrt{5}$

3   The first proportion $\dfrac{AB}{AC} = \dfrac{AC}{CB}$ must be true; the other proportion $\dfrac{AB}{CB} = \dfrac{CB}{AC}$ cannot hold since AB > CB and AC > CB. We see that $\dfrac{AB}{CB} > 1$ and $\dfrac{CB}{AC} < 1$. Hence, it is impossible for $\dfrac{AB}{CB}$ to be equal to $\dfrac{CB}{AC}$.

4   $\dfrac{KL}{FL} = \dfrac{FL}{KF}$ or $\dfrac{FL}{KL} = \dfrac{KF}{FL}$; $FL^2 = KL \cdot KF$

5   The Golden Section

6   **a)** $\dfrac{9}{7}$ **b)** $5\sqrt{3}$ **c)** $\dfrac{8}{3}$ **d)** $4t$ **e)** $\dfrac{3x^2}{2z}$

7   $a = b = c$; If $b \neq c$, then either $b < c$ or $c < b$. Suppose $b < c$. Then $b^2 < bc$, and since b and c are assumed greater than 0, it follows that $b < \sqrt{bc}$. Similarly, $b < c$ implies $bc < c^2$ or $\sqrt{bc} < c$. Hence, $b < \sqrt{bc} < c$. But $a = \sqrt{bc}$, so $b < a < c$.
     The proof for $c < b$ is the same as that given for $b < c$. Simply interchange the b's and c's in the proof.

8   **a)** 4   **b)** 3   **c)** 7   **d)** 2

9　a)　$a^2 = b \cdot c$　　　　b)　$\dfrac{b}{a} = \dfrac{a}{c}$

$a = \dfrac{bc}{a}$　　　　　　$a\left(\dfrac{b}{a}\right) = a\left(\dfrac{a}{c}\right)$

$\dfrac{1}{c}(a) = \dfrac{1}{c}\left(\dfrac{bc}{a}\right)$　　　　$b = \dfrac{a^2}{c}$

$\dfrac{a}{c} = \dfrac{b}{a}$　　　　　　$bc = a^2$

$\dfrac{b}{a} = \dfrac{a}{c}$　　　　　　$a^2 = bc$

SECTION 2 (pages 5–10)

1　Follow the procedure explained on page 6.

2　The Divine Proportion refers to the division of a line segment into extreme and mean ratio. More specifically, it refers to the actual proportion that can be determined if it is known that a point divides a given segment into extreme and mean ratio.

*3　There are, of course, many acceptable ways in which to do this. One such proof is given. (See Figure 2.)

　　1　Draw $\overline{RN}$ and $\overline{NS}$　　　Two points determine a line

2 ΔRNS is a right Δ

An angle inscribed in a semi-circle is a right angle; also, definition of right triangle

3 $\overline{NB}$ is an altitude to the hypotenuse of ΔRNS

$\overline{NB}$ was constructed perpendicular to $\overline{AB}$; also, definition of altitude

4 $\dfrac{RB}{BN} = \dfrac{BN}{BS}$

Theorem 2, page 7

5 $\dfrac{RB}{BN} - 1 = \dfrac{BN}{BS} - 1$

Subtraction property of equality

6 $\dfrac{BN}{BN} = 1$ and $\dfrac{BS}{BS} = 1$

$\dfrac{a}{a} = 1, a \neq 0$

7 $\dfrac{RB}{BN} - \dfrac{BN}{BN} = \dfrac{BN}{BS} - \dfrac{BS}{BS}$

Statements 5 and 6 and substitution property of equality

8 $\dfrac{RB - BN}{BN} = \dfrac{BN - BS}{BS}$

Definition of subtraction of fractions

9 BN = AB and AT = BS

They were constructed this way

10 $\dfrac{RB - AB}{AB} = \dfrac{AB - AT}{AT}$

Statements 8 and 9 and substitution property of equality

11 RB − AB = RA and AB − AT = TB

A is between R and B and T is between A and B

12 $\dfrac{RA}{AB} = \dfrac{TB}{AT}$ Statements 10 and 11 and substitution property of equality

13 $RA = AT$ AT = BS (statement 9); also, M is the midpoint of $\overline{AB}$, thus making BS = RA

14 So $\dfrac{AT}{AB} = \dfrac{TB}{AT}$ Statements 12 and 13 and substitution property of equality

15 $\dfrac{AB}{AT} = \dfrac{AT}{TB}$ If $\dfrac{a}{b} = \dfrac{c}{d}$, where $a \neq 0$ and $c \neq 0$, then $\dfrac{b}{a} = \dfrac{d}{c}$

**4   On a given construction line, say l, mark off a unit segment of convenient length. Call this segment $\overline{AB}$. Using B as an endpoint, mark off on the line l another segment of length five units. Call this segment $\overline{BC}$. Now locate the midpoint of $\overline{AC}$. Call it M. Using M as the center and $\overline{MA}$ or $\overline{MC}$ as the radius, draw a semicircle intersecting line AC. At point B on segment AC, construct a perpendicular to line AC that intersects the semicircle at point T. $\overline{BT}$ is the segment whose length is $\sqrt{5}$. The proof of this is dependent upon Theorem 2 on page 7.

     To generalize, merely make the lengths of segments AB and BC the numbers whose mean proportional you are trying to construct.

SECTION 3 (pages 11–20)

2   See Section 11 for some examples.

3   Equilateral triangle, square, pentagon, hexagon

4   $\dfrac{n(n-3)}{2}$ is the general formula.

5   The relationship is found on page 16. The proportion $\dfrac{OB}{AB} = \dfrac{AB}{AK}$ is true because in Figure 2 point T forms the Golden Section, and Figure 6 was constructed so that OA equals AB of Figure 2 and AB of Figure 6 equals AT of Figure 2.

6   **a)** In order to be a regular pentagon, the five sides must be congruent and the five interior angles must be congruent. Now if the five central angles that we could draw in Figure 6 each have a measure of 72, then $\triangle AOC \cong \triangle COE \cong \triangle EOG \cong \triangle GOI \cong \triangle IOA$ by S.A.S. Hence, AC = CE = EG = GI = IA. In addition, $m \measuredangle OAC + m \measuredangle OAI = m \measuredangle OIA + m \measuredangle OIG = \ldots m \measuredangle ECO + m \measuredangle OCA$. So all interior angles of ACEGI are congruent. Thus, ACEGI is regular.

   b) $\measuredangle BOA$ is a central angle. By showing $m \measuredangle BOA = 36$, we can establish in a manner similar to that in part a that ABCDEFGHIJ is a regular decagon. Hence, each of its central angles is 36, and it follows then that the sum of any two angles would be 72. That ACEGI is a regular pentagon follows immediately from an argument as in part a.

c) Since $\triangle OBA \sim \triangle BAK$ and since $\triangle OBA$ is isosceles (two of its sides are radial segments), it follows that $\triangle BAK$ is isosceles. Hence, BK = BA. Since OK = BA, then OK = BK. Consequently, $\triangle OKB$ is isosceles.

d) In $\triangle OBA$ we know that $m \angle 4 + m \angle 1 + m \angle 2 = 180$. Also, we can observe in Figure 5 that $m \angle 1 = m \angle 5 + m \angle 3$. This equation becomes $m \angle 1 = m \angle 4 + m \angle 4 = 2(m \angle 4)$, because $m \angle 5 = m \angle 4$ and $m \angle 3 = m \angle 4$. Since we also know that $m \angle 1 = m \angle 2$, it follows that $m \angle 2 = 2(m \angle 4)$. Returning to the equation in the first line of this proof and substituting, we get, $m \angle 4 + 2(m \angle 4) + 2(m \angle 4) = 180$, or $5 (m \angle 4) = 180$.

SECTION 4 (pages 21–26)

1 **a)** If $\dfrac{a}{b} = \dfrac{c}{d}$, then $\left(\dfrac{a}{b}\right)\left(\dfrac{db}{ac}\right) = \left(\dfrac{c}{d}\right)\left(\dfrac{db}{ac}\right)$. Simplifying, we get $\dfrac{d}{c} = \dfrac{b}{a}$, or $\dfrac{b}{a} = \dfrac{d}{c}$. Therefore, if $\dfrac{a}{b} = \dfrac{c}{d}$, then $\dfrac{b}{a} = \dfrac{d}{c}$.

b) If $\dfrac{a}{b} = \dfrac{c}{d}$, then $\dfrac{a}{b} + 1 = \dfrac{c}{d} + 1$. But $1 = \dfrac{b}{b}$ and $1 = \dfrac{d}{d}$, so $\dfrac{a}{b} + \dfrac{b}{b} = \dfrac{c}{d} + \dfrac{d}{d}$ or $\dfrac{a+b}{b} = \dfrac{c+d}{d}$. Hence, if $\dfrac{a}{b} = \dfrac{c}{d}$, then $\dfrac{a+b}{b} = \dfrac{c+d}{d}$.

c) If $\dfrac{a}{b} = \dfrac{c}{d}$, then $\left(\dfrac{a}{b}\right)\left(\dfrac{b}{c}\right) = \left(\dfrac{c}{d}\right)\left(\dfrac{b}{c}\right)$. Simplifying, we have $\dfrac{a}{c} = \dfrac{b}{d}$. Therefore, if $\dfrac{a}{b} = \dfrac{c}{d}$, then $\dfrac{a}{c} = \dfrac{b}{d}$.

d) If $\frac{a}{b} = \frac{c}{d}$, then $\left(\frac{a}{b}\right)\left(\frac{d}{a}\right) = \left(\frac{c}{d}\right)\left(\frac{d}{a}\right)$. Simplifying, we get

$\frac{d}{b} = \frac{c}{a}$. Hence, if $\frac{a}{b} = \frac{c}{d}$, then $\frac{d}{b} = \frac{c}{a}$.

e) If $\frac{a}{b} = \frac{c}{d}$, then $\frac{a}{b} - 1 = \frac{c}{d} - 1$. But since $\frac{b}{b} = 1$ and

$\frac{d}{d} = 1$, we know that $\frac{a}{b} - \frac{b}{b} = \frac{c}{d} - \frac{d}{d}$. Simplifying, we

have $\frac{a-b}{b} = \frac{c-d}{d}$. Therefore, if $\frac{a}{b} = \frac{c}{d}$, then

$\frac{a-b}{b} = \frac{c-d}{d}$.

*2   Suppose AB = 3. Let AC = x, so that CB = $3 - x$. Then

from $\frac{AC}{AB} = \frac{CB}{AC}$, we can obtain $\frac{x}{3} = \frac{3-x}{x}$, or $x^2 = 9 - 3x$,

or $x^2 + 3x - 9 = 0$. So $x = \frac{-3 \pm \sqrt{9 + 36}}{2} = \frac{-3 \pm \sqrt{45}}{2} =$

$\frac{-3 \pm 3\sqrt{5}}{2}$. Considering only the positive root, we have

$x = \frac{-3 + 3\sqrt{5}}{2}$. Hence $\frac{AC}{AB} = \frac{x}{3} = \frac{\left(\dfrac{-3 + 3\sqrt{5}}{2}\right)}{3} = \frac{-1 + \sqrt{5}}{2}$.

To obtain the generalization of this argument,
suppose we let AB = y where y is any positive real
number. Then we see, as in the first part of this
problem, that $\frac{AC}{AB} = \frac{CB}{AC}$ becomes $\frac{AC}{y} = \frac{y - AC}{AC}$ or

$AC^2 = y^2 - yAC$. So $AC^2 + yAC - y^2 = 0$. Hence, AC =

$\frac{-y \pm \sqrt{y^2 + 4y^2}}{2}$. Since y is a positive real number, we

get AC = $\frac{-y \pm y\sqrt{5}}{2}$. Discarding the negative root as before, we have AC $= \dfrac{-y + y\sqrt{5}}{2}$. Therefore, the ratio of

$\dfrac{AC}{AB}$ becomes $\dfrac{\left(\dfrac{-y + y\sqrt{5}}{2}\right)}{y}$ or simply $\dfrac{-1 + \sqrt{5}}{2}$.

3   $\dfrac{1}{x} = \dfrac{x}{1 - x}$ Definition of the Golden Section

     $1 - x = x^2$ Multiplying both sides of the equation by $x(1 - x)$

     $0 = x^2 + x - 1$ Adding $x - 1$ to both sides of the equation

     $x^2 + x - 1 = 0$ Symmetric property of equality

     $x = \dfrac{-1 \pm \sqrt{5}}{2}$ Quadratic formula

4   Completing the square

$$x^2 + x - 1 = 0$$
$$x^2 + x = 1$$
$$x^2 + x + \frac{1}{4} = 1 + \frac{1}{4}$$
$$\left(x + \frac{1}{2}\right)^2 = \frac{5}{4}$$
$$x + \frac{1}{2} = \pm \sqrt{\frac{5}{4}}$$
$$x = -\frac{1}{2} \pm \frac{\sqrt{5}}{2}$$

$$x = \frac{-1 \pm \sqrt{5}}{2}$$

Using the quadratic formula

$$x = \frac{-b \pm \sqrt{b^2 - 4ac}}{2 \cdot a}$$

$$x = \frac{-1 \pm \sqrt{1 - 4 \cdot 1 \cdot (-1)}}{2 \cdot 1}$$

$$x = \frac{-1 \pm \sqrt{5}}{2}$$

5 $\tau - 1 = \dfrac{1}{\tau}$

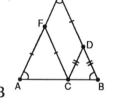

7 *Show:* $\dfrac{AB}{AC} = \dfrac{AC}{CB}$

Given: $\dfrac{AE}{AF} = \dfrac{AF}{FE}$ and $\dfrac{EB}{ED} = \dfrac{ED}{DB}$

AF = FC, CD = DB, and AE = EB

We know that $\dfrac{AE}{AF} = \dfrac{AF}{FE} = \tau$ and that $\dfrac{EB}{ED} = \dfrac{ED}{DB} = \tau$,

so it follows that $\dfrac{AE}{AF} = \dfrac{EB}{ED}$. Since the numerators of these fractions are equal, the denominators must be equal. Hence, AF = ED. We know also that AF = FC (this is given), so it follows that <u>FC = ED</u>. Similarly, $\dfrac{AF}{FE} = \dfrac{ED}{DB}$. We established, however, that AF = ED; therefore, if the numerators of two equal fractions are equal, so are their denominators. Hence, FE = DB. As before, we are given that CD = DB. Substitution then yields <u>FE = CD</u>. The underlined statements indicate that the opposite sides of quadrilateral FCDE are congruent. Therefore, it follows that FCDE is a parallelogram. Knowing this, it is fairly easy to see that $m \angle AFC = m \angle AED$ and that $m \angle AEB = m \angle CDB$. Observing that angle A is shared by $\triangle AEB$ and $\triangle AFC$ and that angle B is shared by $\triangle AEB$ and $\triangle CDB$, we see that $\triangle AEB \sim \triangle AFC$ and $\triangle AEB \sim \triangle CDB$ (by the angle-angle similarity theorem). Hence, $\triangle AFC \sim \triangle CDB$ (two triangles similar to the same triangle are similar to each other). From the above similarities between the three triangles, it follows from the definition of similarity that $\dfrac{AB}{AC} = \dfrac{AC}{CB}$.

SECTION 5 (pages 27–32)

1   The three cases to consider are illustrated below:

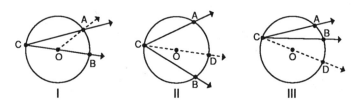

Following are sketches of the proofs:

Case I: The center of the circle is on one of the sides of the inscribed angle. Draw $\overrightarrow{OA}$. Then $AO = CO$ and $m \angle CAO = m \angle ACO$. Since $m \angle AOB = m \angle ACO + m \angle CAO$, it follows that $m \angle ACO = \dfrac{1}{2} m \angle AOB$. But $\angle AOB$ is a central angle, so $m \angle AOB = m \overset{\frown}{AB}$. Hence, $m \angle ACO = \dfrac{1}{2} m \overset{\frown}{AB}$.

Case II: The center of the circle is inside the angle. Draw $\overrightarrow{CO}$ and call its point of intersection with the circle D. From Case I, $m \sphericalangle ACO = \frac{1}{2} m \overparen{AD}$ and

$m \sphericalangle OCB = \frac{1}{2} m \overparen{DB}$. Adding, we have $m \sphericalangle ACO +$

$m \sphericalangle OCB = \frac{1}{2}(m \overparen{AD} + m \overparen{DB})$. But $m \sphericalangle ACO + m \sphericalangle OCB$

$= m \sphericalangle ACB$ and $m \overparen{AD} + m \overparen{DB} = m \overparen{ADB}$.

Substituting, we have $m \sphericalangle ACB = \frac{1}{2} m \overparen{AOB}$.

Case III: The center of the circle is outside of $\sphericalangle ACB$. Draw $\overrightarrow{CO}$ intersecting the circle at D. $m \sphericalangle ACO =$

$\frac{1}{2} m \overparen{ABD}$ and $m \sphericalangle BCO = \frac{1}{2} m \overparen{BD}$. Subtracting, we

have $m \sphericalangle ACO - m \sphericalangle BCO = \frac{1}{2}(m \overparen{ABD} - m \overparen{BD})$.

However, $m \sphericalangle ACO - m \sphericalangle BCO = m \sphericalangle ACB$ and

$m \overparen{ABD} - m \overparen{BD} = m \overparen{AB}$. Substituting, we have

$m \sphericalangle ACB = \frac{1}{2} m \overparen{AB}$.

2 *Given:* Circle O with $\overparen{AB}$ and $\overparen{CD}$ having equal measures

Prove: AB = CD

Draw $\overline{AO}$, $\overline{OB}$, $\overline{OC}$, and $\overline{OD}$. Being radial segments of the same circle, AO = OB = OC = OD. Now $m \overset{\frown}{AB} = m \overset{\frown}{CD}$ means $m \sphericalangle AOB = m \sphericalangle COD$. Hence, $\triangle AOB \cong \triangle COD$ (S.A.S.). It follows that AB = CD.

3   *Given:* Circle O with AB = CD

   Prove: $m \overset{\frown}{AB} = m \overset{\frown}{CD}$

Draw $\overline{OA}$, $\overline{OB}$, $\overline{OC}$, and $\overline{OD}$. As in Problem 2, AO = OB = OC = OD. Since AB = CD, $\triangle AOB \cong \triangle DOC$ (S.S.S.). So $m \sphericalangle AOB = m \sphericalangle DOC$. Hence, $m \overset{\frown}{AB} = m \overset{\frown}{CD}$.

4   In Figure 8 we are given that ABCDE is a regular pentagon. This means that AB = BC = CD = DE = EA. Hence, by making use of Theorem 7, we know that $m \overset{\frown}{AB} = m \overset{\frown}{BC} = m \overset{\frown}{CD} = m \overset{\frown}{DE} = m \overset{\frown}{EA}$. Since each of these arcs has the same measure, we know that the sum of any two arcs will equal the sum of any other two. Hence, for example, $m \overset{\frown}{AB} + m \overset{\frown}{BC} = m \overset{\frown}{BC} + m \overset{\frown}{CD}$. Since $m \overset{\frown}{AB} + m \overset{\frown}{BC} = m \overset{\frown}{AC}$ and $m \overset{\frown}{BC} + m \overset{\frown}{CD} = m \overset{\frown}{BD}$, upon substituting in the above equation, we get $m \overset{\frown}{AC} = m \overset{\frown}{BD}$. Now, using Theorem 6, we have AC = BD. In a similar manner, we can show that all other pairs of diagonals of ABCDE are congruent.

*5   Here is an outline of a proof. Recall that in order to show FGHIJ is a regular pentagon, we must show two things: first, its sides are all congruent and second, its interior angles are all equal in measure.

To do the first, merely prove that the five triangles CHB, BGA, AFE, EJD, and DIC are congruent. This is done by noting that AB = BC = CD = DE = EA and then using the A.S.A. congruence theorem. After we have shown these triangles to be congruent, we use the fact that AG = GB = BH = . . . = AF (because they are all legs of congruent isosceles triangles) along with the S.A.S. congruence theorem, to prove that triangles AGF, BHG, CIH, DJI, and EFJ are congruent. Having done this, we can easily conclude that FG = GH = HI = IJ = JF. This takes care of the first part.

We have already established in part one of this problem that the five triangles CHB, BGA, AFE, EJD, and DIC are congruent. Hence, $m \sphericalangle CHB = m \sphericalangle BGA = m \sphericalangle AFE = m \sphericalangle EJD = m \sphericalangle DIC$. Since each one of the interior angles of FGHIJ is a vertical angle to some one of the above five, it follows that they must be equal.

Hence, from the above arguments, we may conclude that FGHIJ is a regular pentagon.

6   $\dfrac{(n-2)\,180}{n}$; no, this formula does not apply to polygons that are not regular, because the interior angles would not all be equal in measure.

7   $(n-2)180$; yes, this formula is applicable to polygons that are not regular.

*8   Suppose that in △ABC
      $m \angle B = m \angle C$. To show that △ABC
      is isosceles, we need to show
      that AB = AC. To do this, drop
      a perpendicular to $\overline{BC}$ from A.
      Call it $\overline{AM}$. Certainly AM = AM;
      $m \angle AMC = m \angle BMA$, since both
      are right angles. Hence, since
      we are given $m \angle B = m \angle C$,
      △AMB ≅ △AMC by S.A.A. Thus,
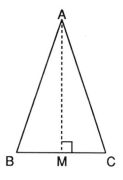
the corresponding sides $\overline{AB}$ and $\overline{AC}$ are congruent, so
△ABC is isosceles.

Project

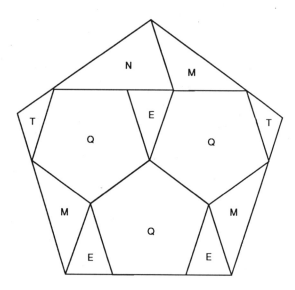

SECTION 6 (pages 33–42)

1 **a)** 1, 3, 9, 27, 81, 243, 729
   b) $\sqrt{3}, 2\sqrt{3}, 4\sqrt{3}, 8\sqrt{3}, 16\sqrt{3}, 32\sqrt{3}, 64\sqrt{3}$
   c) $\frac{1}{5}$, 1, 5, 25, 125, 625, 3125
   d) .1, $1 \times 10^{-3}, 1 \times 10^{-5}, 1 \times 10^{-7}, 1 \times 10^{-9}, 1 \times 10^{-11},$
     1×10^{-13}

2 **a)** Not geometric
   b) Geometric; $r = \frac{1}{2}$
   c) Geometric; $r = \frac{\sqrt{2}}{2}$
   d) Not geometric

Sketches or outlines of the proofs for Problems 3 through 8 follow. Students probably will give more detailed proofs.

3 AE = ED = DC = CB = BA. $m \angle FEA = m \angle FAE = m \angle GAB = m \angle GBA = \ldots = m \angle JDE = m \angle JED = 36$. Hence, the five triangles are congruent by A.S.A.

4 By Problem 3, JE = EF = FA = AG = $\ldots$ = ID = JD. Also, $m \angle JEF = m \angle FAG = m \angle GBH = m \angle HCI = m \angle IDJ = 36$. Therefore, the five triangles are congruent by S.A.S.

5 $m \angle GEA = m \angle IED = 36$. $m \angle GAE = m \angle IDE = 72$. Since DE = EA, $\triangle EDI \cong \triangle EAG$ (A.S.A.). The remaining eight triangles can be proved congruent in a similar manner.

6 From Problem 4 we can show that all sides of FGHIJ
   are of equal length. Also, each of the isosceles triangles
   in Problem 4 has base angles whose measures are 72.
   So, for example, in ∆AGF, $m \angle$AGF = $m \angle$AFG = 72.
   Hence, $m \angle$HGF = $m \angle$JFG = 108. Similarly, $m \angle$GHI =
   $m \angle$HIJ = $m \angle$IJF = 108. It follows that FGHIJ is a
   regular pentagon.

7 In Problem 6 we proved that FGHIJ is a regular
   pentagon. Therefore, $m \angle$HJF = $m \angle$HFJ = 72. But
   $m \angle$FJE = $m \angle$JFE = 72. So, by A.S.A., ∆EFJ ≅ ∆HFJ.
   In a similar manner we can show that ∆AGF ≅ ∆IGF,
   ∆BHG ≅ ∆JHG, ∆CIH ≅ ∆FIH, and ∆DJI ≅ ∆GJI. Hence,
   all 10 triangles would be congruent.

8 The five rhombi are DEAH, EABI, AJCB, BFDC, and
   CGED. To prove that these quadrilaterals are rhombi,
   we note that in Problem 5 we proved ten triangles
   congruent. Since it is easy to show that these triangles
   are isosceles, it follows that all segments making up
   the sides of the five quadrilaterals are of equal length.
   Hence, the above quadrilaterals are parallelograms
   and also rhombi.
       The smaller rhombi are EFHJ, AGIF, BHJG, CIFH,
   and DJGI; the smallest rhombi are JFGL, FGHM,
   GHIN, HIJO, and IJFK. That the first five are rhombi
   follows from Problem 7 and the fact that all triangles
   in Problem 7 are isosceles. Hence, for example, FE =
   EJ = JH = HF. So EFHJ is a rhombus. A similar proof
   can be given for AGIF, BHJG, CIFH, and DJGI.
       Since FGHIJ was proved to be a regular pentagon,
   the proof that the five smallest quadrilaterals (JFGL,
   FGHM, GHIN, HIJO, and IJFK) are rhombi is handled

in the same way as for the rhombi of the large pentagon ABCDE in the first part of this problem.

9   In the section we showed that $\frac{a}{b} = \frac{b}{c} = \frac{c}{d} = \frac{d}{e} = \frac{e}{f}$. By proportion inversion, therefore, $\frac{f}{e} = \frac{e}{d} = \frac{d}{c} = \frac{c}{b} = \frac{b}{a}$. It follows that f, e, d, c, b, a is a geometric progression. Looking at Figure 9 and using the fact that JF = JO = d and OF = JN = e, we can write JN + NO = JO; or $e + f = d$. But $e = fr$ and $d = fr^2$, where r is the common ratio of the progression. Consequently, $fr + f = fr^2$ or $r + 1 = r^2$ or $r^2 - r - 1 = 0$. Solving the quadratic equation $r^2 - r - 1 = 0$ for r yields
$$r = \frac{1 + \sqrt{5}}{2} = \tau.$$

*10   We note the following:

The shaded area of I = 5 × the area of ΔABF.
The shaded area of II = 5 × the area of ΔABG.
The shaded area of III = 5 × the area of ΔAFG.

We will show that the areas of ΔABF, ΔABG, and ΔAFG form a geometric progression with ratio $\frac{1}{\tau}$. Then it will follow that the shaded areas of I, II, and III form a geometric progression with ratio $\frac{1}{\tau}$, because multiplying the areas of ΔABF, ΔABG, and ΔAFG by 5 will not affect the *ratio* of the areas. For example, a, ar, ar^2 is a geometric progression with ratio r. Multiplying each element by 5 yields the geometric progression $5a$, $5ar$, $5ar^2$, with the same ratio, r.

The areas of $\triangle ABF$, $\triangle ABG$, and $\triangle AFG$ are given below.

$$\text{Area of } \triangle ABF = \frac{1}{2} FB \cdot h$$

$$\text{Area of } \triangle ABG = \frac{1}{2} GB \cdot h = \frac{1}{2} FB \cdot \frac{1}{\tau} h$$

$$\text{Area of } \triangle AFG = \frac{1}{2} FG \cdot h = \frac{1}{2} FB \cdot \left(\frac{1}{\tau}\right)^2 h$$

Each of the triangles has the same altitude, which we have denoted by h. From our work in this lesson, we know that $GB = FB \cdot \frac{1}{\tau}$ and $FG = FB \cdot \left(\frac{1}{\tau}\right)^2$. Hence, by substituting these values for GB and FG, we obtained $\frac{1}{2} FB \cdot \frac{1}{\tau} h$ as the area of $\triangle ABG$ and $\frac{1}{2} FB \cdot \left(\frac{1}{\tau}\right)^2 h$ as the area of $\triangle AFG$. It is easy to see that the areas of $\triangle ABF$, $\triangle ABG$, and $\triangle AFG$ do form a geometric progression with a common ratio of $\frac{1}{\tau}$.

11 $DE = \frac{1}{\tau}$, $EF = \left(\frac{1}{\tau}\right)^2$, $FG = \left(\frac{1}{\tau}\right)^3$, $GH = \left(\frac{1}{\tau}\right)^4$, $HI = \left(\frac{1}{\tau}\right)^5$, $IJ = \left(\frac{1}{\tau}\right)^6$, and $JK = \left(\frac{1}{\tau}\right)^7$.

12 $\tau, \tau^2, \tau^3, \tau^4, \tau^5$

**13 Bisect ⟩MNO and ⟩MON. Call these bisectors $\overline{ND}$ and $\overline{OT}$. Draw in the other segments as shown below. MDONT is the required pentagon.

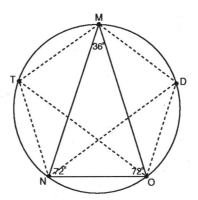

    To prove that the construction procedure is correct, we note that we bisected ⟩MNO and ⟩MON, obtaining four angles each of whose measure is 36. Therefore, m ⟩NMO = m ⟩MND = m ⟩DNO = m ⟩TON = m ⟩TOM. It follows that MD = DO = ON = NT = TM. (Why?) Therefore, pentagon MDONT has equal sides. All that remains is to show that m ⟩MDO = m ⟩DON = m ⟩ONT = m ⟩NTM = m ⟩TMD. To do this we notice that m $\overarc{TN}$ = m $\overarc{MD}$ (Theorem 7, page 28). Hence, m $\overarc{TNOD}$ = m $\overarc{MDON}$. (Why?) It follows that m ⟩TMD = m ⟩MTN. In a similar manner we can prove that the remaining interior angles of MDONT are congruent, and thus the pentagon is regular. This completes the proof.

*14 $CD = \tau$; $EG = \dfrac{1}{\tau}$; $EF = \dfrac{1}{\tau^2}$; $FH = \dfrac{1}{\tau^3}$; $\overline{HQ}$ is a diagonal of the small pentagram. Hence, its length is τ times the length of one of its sides, say $\overline{HM}$. But HM = 2HF, or $2\left(\dfrac{1}{\tau^3}\right)$. So $HQ = \tau HM = \tau \cdot 2\left(\dfrac{1}{\tau^3}\right) = \dfrac{2}{\tau^2}$.

15   The sum of the series being sought (*i.e.*, $2 + 2\left(\dfrac{1}{\tau}\right) + 2\left(\dfrac{1}{\tau^2}\right) + 2\left(\dfrac{1}{\tau^3}\right) + 2\left(\dfrac{1}{\tau^4}\right) + 2\left(\dfrac{1}{\tau^5}\right) + \ldots$) is $\dfrac{2}{1 - \left(\dfrac{1}{\tau}\right)} \approx 5.24$.

SECTION 7 (pages 43–60)

1   Two definitions are:
   a) A Golden Rectangle is a rectangle for which the ratio of its longer side to its shorter side is equal to τ.
   b) A Golden Rectangle is one for which the ratio of the shorter to the longer side is $\dfrac{1}{\tau}$.

2   To do this construction, simply use $\overline{GR}$ in the same way $\overline{AB}$ was used in Figure 11.

3   $OB = \sqrt{2}$, $OC = \sqrt{3}$, $OD = 2$, $OE = \sqrt{5}$, $OF = \sqrt{6}$, $OG = \sqrt{7}$

4   $AE = \sqrt{2}$, $AF = \sqrt{3}$, $AG = 2$, $AH = \sqrt{5}$, $AI = \sqrt{6}$, $AJ = \sqrt{7}$

*5 **a)** Both $\dfrac{CB}{EC}$ and $\dfrac{DC}{DA}$ are equal to τ, or $\dfrac{1 + \sqrt{5}}{2}$, because ABCD is a Golden Rectangle and ECBF is its reciprocal, which means, of course, that ABCD ~ ECBF.

    b) This follows immediately from the definition of Golden Rectangle.

    c) From part b, $\dfrac{x}{y} = \dfrac{x + y}{x}$. Since $\dfrac{x + y}{x} = \dfrac{x}{x} + \dfrac{y}{x} = 1 + \dfrac{y}{x}$, it follows that $\dfrac{x}{y} = 1 + \dfrac{y}{x}$.

    d) This follows from part c and the multiplication property of equality.

    e) This is an immediate consequence of part d. The equation looks very much like the characteristic equation we encountered in Figure 7b.

    f) τ

    g) $\dfrac{x}{y} = \tau = \dfrac{1 + \sqrt{5}}{2}$

    h) $\dfrac{x}{y}$ represents the ratio of the longer side to the shorter side of ECBF. This means ECBF is a Golden Rectangle.

  6 The shorter side of each rectangle is found from the preceding rectangle by subtracting its shorter side from its longer side. The longer side of each rectangle is just the shorter side of the preceding rectangle. The pattern in the coefficients of m and n is the Fibonacci sequence.

*7 Although there are many Golden Rectangles that can be formed, all of them can be placed into five categories. They can be formed from any of the segments of the lengths below.

1) AB and BE
2) BA and BG
3) AG and GF
4) GF and OF
5) OF and ON

This follows from the work in Section 6 and from Problem 1 of this section.

8   Merely draw in a diagonal of your rectangles, and then from one of the other vertices construct a line perpendicular to the diagonal. Below, for example, FBCE is a reciprocal of ABCD.

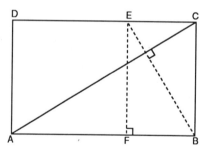

*9   **a)** AEFD is a Golden Rectangle, so that $\dfrac{AE}{AD} = \dfrac{AE}{1} = AE$

= τ. Now AE − AB = τ − 1 = $\dfrac{1}{\tau}$ by Problem 5 of

Section 4. Hence, $\dfrac{FE}{BE} = \dfrac{1}{\dfrac{1}{\tau}} = \tau$. So $\dfrac{AE}{FE} = \dfrac{FE}{FC} = \tau$, and

$\triangle$AEF ~ $\triangle$EFC by S.A.S. Thus $m \not{\angle}$FAE = $m \not{\angle}$FEC, and $m \angle AFE = m \angle AFE$. Since, therefore, two angles of $\triangle AEF$ are congruent to two angles of $\triangle EQF$, it follows that $m \angle FQE = m \angle AEF = 90$.

Thus, $\overline{AF} \perp \overline{EC}$.

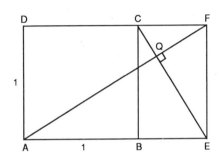

b) In order to show that ABCD is a square, all we need to do is to show that AD = AB. Now from part a, we know $\triangle AEF \sim \triangle EFC$. Hence, $\dfrac{AE}{EF} = \dfrac{EF}{FC}$. But

FC = EB, so we have that $\dfrac{AE}{EF} = \dfrac{EF}{EB}$. Since AEFD is

a Golden Rectangle, $\dfrac{AE}{AD} = \dfrac{AD}{AE - AD}$. However, EF

= AD, so it follows from the two proportions that EB = AE − AD. From the diagram, however, EB also equals AE − AB. It follows then that AD = AB. Hence ABCD is a square.

10   This results as a direct consequence of Problem 9b.

*11   Merely follow the directions for constructing Golden Rectangle AEDB as given in Figure 11. The perimeter of AEDB is $1 + 1 + 2 \cdot \dfrac{1}{\tau} = 1 + 1 + (-1 + \sqrt{5}) = 1 + \sqrt{5}$.

12 **a) $\triangle ACB \sim \triangle BCT$; A.A.

b) $\overline{AT}$, $\overline{TB}$, and $\overline{BC}$

c) This is an immediate consequence of part a and the fact that $TC = a - AT = a - b$.

d) From part c, $\dfrac{a}{b} = \dfrac{b}{a-b}$, or $a^2 - ab = b^2$.

So $\dfrac{a^2}{b^2} - \dfrac{ab}{b^2} = 1$. Therefore, $\left(\dfrac{a}{b}\right)^2 - \dfrac{a}{b} = 1$ and

$\left(\dfrac{a}{b}\right)^2 - \dfrac{a}{b} - 1 = 0$. The equation will look familiar if

you did part e of Problem 5 in this section.

e) $\dfrac{a}{b} = \dfrac{1 + \sqrt{5}}{2} = \tau$

f) It represents the ratio of the length of a leg of a 36-72-72 triangle to the base. Part e tells us that, according to the definition of Golden Triangle, any 36-72-72 triangle is a Golden Triangle.

g) In order to generate smaller and smaller Golden Triangles, you could bisect ∢BCA to form $\triangle CTR$, which is similar to $\triangle ABC$. Then bisect ∢RTC to form $\triangle TRS$, which is similar to $\triangle ABC$. Then bisect ∢TRS to form $\triangle RSU$, which is similar to $\triangle ABC$, and so on.

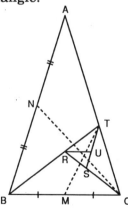

h) No; it is only with these angles that we can obtain the characteristic equation of part d that leads to the ratio of a side to the base being τ.

**13 From part h of Problem 12, $\triangle ABC$ must be a 36-72-72 triangle. Since $\triangle ACB \sim \triangle BDC$, $\triangle BDC$ also is a 36-72-72 triangle. We know that $\triangle ADB$ is isosceles, so AD = DB. Now since BD = BC, it follows that AD = BC. But from Problem 12, $\dfrac{AC}{BC} = \tau$. So $\dfrac{AC}{AD} = \tau$.

Now the area of $\triangle ABC = \dfrac{1}{2} \cdot AC \cdot BG$ and the area of $\triangle ADB = \dfrac{1}{2} \cdot AD \cdot BG$. Therefore, $\dfrac{\text{area of } \triangle ABC}{\text{area of } \triangle ADB}$

$= \dfrac{\dfrac{1}{2} \cdot AC \cdot BG}{\dfrac{1}{2} \cdot AD \cdot BG} = \dfrac{AC}{AD}$. Since $\dfrac{AC}{AD} = \tau$, then

$\dfrac{\text{area of } \triangle ABC}{\text{area of } \triangle ADB} = \tau$.

14 Since $\overline{AB} \parallel \overline{DC}$, $m \measuredangle CAB = m \measuredangle ACD$. In $\triangle GBC$ and $\triangle GMC$, $\measuredangle GMC$ and $\measuredangle GCB$ are right angles. $\measuredangle CGM$ is common to both triangles. Hence, $m \measuredangle GCM = m \measuredangle CBM$. So $m \measuredangle CAB = m \measuredangle CBM$. Since $\measuredangle GCB$ and $\measuredangle CBA$ are right angles, $\triangle ABC \sim \triangle BCG$. (By A.A.) Hence, $\dfrac{GC}{CB} = \dfrac{CB}{AB}$. Since we are given that both ABCD and GNBC are rectangles, and since we have just shown that their corresponding sides are proportional, it follows that ABCD $\sim$ GNBC.

NOTE: If one accepts the statement on page 46 that "the reciprocal of a rectangle is another rectangle smaller is size but *similar* in shape to the original . . . ," as a definition, then the proof of this problem is trivial.

**15 From our work in Section 4 with the characteristic equation, we know $\left(\frac{1}{\tau}\right)^2 + \frac{1}{\tau} - 1 = 0$, or $\left(\frac{1}{\tau}\right)^2 = 1 - \frac{1}{\tau}$.

b) The formula for finding the sum of an infinite geometric progression is given by $\frac{a}{1-r}$. (What this formula actually does is to give us the limit of the sum.) In this formula a is the first term of the progression and r is the common ratio. Referring to the figure in this problem, we note that AE = $\sqrt{2}$. To get EF, we need the length of one of the sides of the square having EF as diagonal. But $\frac{DC}{DA} = \tau$ or $\frac{(1 + EC)}{1} = \tau$ or $1 + EC = \tau$ or $EC = \tau - 1$. By Problem 5 of Section 4, however, $\tau - 1 = \frac{1}{\tau}$. Hence, $EC = \frac{1}{\tau}$. So EF $= \left(\frac{1}{\tau}\right)\sqrt{2}$. Continuing, GF = BF $\sqrt{2}$.

But $\frac{LB}{BF} = \tau$, or $\frac{\left(\frac{1}{\tau}\right)}{BF} = \tau$, or BF $= \left(\frac{1}{\tau}\right)^2$. So GF $= \left(\frac{1}{\tau}\right)^2 \sqrt{2}$.

Similarly, GH = GL $\sqrt{2}$. But $\frac{LM}{GL} = \tau$, or $\frac{\left(\frac{1}{\tau}\right)^2}{GL} = \tau$, or

$GL = \left(\dfrac{1}{\tau}\right)^3$. Therefore, $GH = \left(\dfrac{1}{\tau}\right)^3 \sqrt{2}$. Continuing in

this manner, $HI = \left(\dfrac{1}{\tau}\right)^4 \sqrt{2}$, $IJ = \left(\dfrac{1}{\tau}\right)^5 \sqrt{2}$, $JK = $

$\left(\dfrac{1}{\tau}\right)^6 \sqrt{2}$, and so on. Thus, the progression of lengths

AE, EF, FG, GH, HI, ... becomes $\sqrt{2}$, $\left(\dfrac{1}{\tau}\right)\sqrt{2}$,

$\left(\dfrac{1}{\tau}\right)^2 \sqrt{2}$, $\left(\dfrac{1}{\tau}\right)^3 \sqrt{2}$, $\left(\dfrac{1}{\tau}\right)^4 \sqrt{2}$, ... This is an infinite

geometric progression with $a = \sqrt{2}$ and $r = \dfrac{1}{\tau}$. Hence,

to find the limit of the sum of this infinite

progression, we use the formula $\dfrac{a}{1-r}$. Substituting,

we get $\dfrac{\sqrt{2}}{1 - \left(\dfrac{1}{\tau}\right)}$. From part a of this problem,

however, $1 - \dfrac{1}{\tau} = \left(\dfrac{1}{\tau}\right)^2$. Consequently, the sum is

$\dfrac{\sqrt{2}}{\left(\dfrac{1}{\tau}\right)^2}$, or $\tau^2 \sqrt{2}$, which is what we wished to show.

c)   3.70

e)   No, but it will become increasingly close to $\tau^2\sqrt{2}$.

**16   Since the point that divides a segment into extreme
and mean ratio is unique, and since ABCD is a square,
it follows that HB = BE and EC = CF. With this in
mind, here is one proof.

$$\frac{BC}{BE} = \frac{BE}{EC}$$

$$\frac{BC^2}{BE^2} = \frac{BE^2}{EC^2}$$

$$\frac{(BE + EC)^2}{BE^2} = \frac{BE^2}{EC^2}$$

$$\frac{(BE^2 + 2BE \cdot EC + EC^2)}{BE^2} = \frac{BE^2}{EC^2}$$

$$\frac{(2BE^2 + 2(2 \cdot BE \cdot EC) + 2EC^2)}{2BE^2} = \frac{2BE^2}{2EC^2}$$

$$\frac{(HE^2 + 2(2 \cdot BE \cdot EC) + FE^2)}{HE^2} = \frac{HE^2}{FE^2}$$

(Since $2BE^2 = BE^2 + BE^2 = HE^2$ and $2EC^2 = EC^2 + EC^2 = FE^2$)

$$\frac{(HE^2 + 2\sqrt{2BE^2}\sqrt{2EC^2} + FE^2)}{HE^2} = \frac{HE^2}{FE^2}$$

$$\frac{(HE^2 + 2HE \cdot FE + FE^2)}{HE^2} = \frac{HE^2}{FE^2} \text{ (by the Pythagorean}$$
theorem)

$$\frac{(HE + FE)^2}{HE^2} = \frac{HE^2}{FE^2}$$

Since all the numbers we are dealing with are positive numbers, it follows:

$$\frac{HE + FE}{HE} = \frac{HE}{FE}.$$

Therefore, rectangle GHEF is a Golden Rectangle.

**17 a) We will show that the length of $\overline{BC}$ in the inscribed regular pentagon below is equal to $\frac{R}{2}\sqrt{10 - 2\sqrt{5}}$.

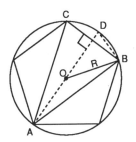

$\triangle ABC$ and $\triangle OBD$ are both Golden Triangles — for a definition of Golden Triangle, see Section 7, Problem 12. Hence, $\frac{AB}{BC} = \frac{R}{BD}$. Also, $\frac{BC}{AB} = \frac{BD}{R}$, or

$BC = \frac{AB \cdot BD}{R}$. But $\triangle ABD$ is a right triangle

since $\overline{AD}$ is a diametral chord. Hence, BD =

$\sqrt{AD^2 - AB^2} = \sqrt{(2R)^2 - AB^2}$. $\overline{AB}$ is a side of a pentagram; therefore, from Section 6 we know that AB = BC · τ. When we substitute, we find that BC = $\dfrac{\tau \cdot BC}{R} \sqrt{(2R)^2 - (\tau \cdot BC)^2}$.

To complete the derivation, we perform the calculations that follow.

$$1 = \frac{\tau}{R} \sqrt{4R^2 - \tau^2 (BC)^2}$$

$$\left(\frac{R}{\tau}\right)^2 = 4R^2 - \tau^2 (BC)^2$$

$$\frac{R^2}{\tau^2} - 4R^2 = -\tau^2 (BC)^2$$

$$\sqrt{\frac{1}{\tau^2}\left(4R^2 - \frac{R^2}{\tau^2}\right)} = BC$$

$$\frac{R}{\tau}\sqrt{4 - \frac{1}{\tau^2}} = BC$$

$$\frac{R}{\tau}\sqrt{4 - \frac{1}{\left(\dfrac{1 + \sqrt{5}}{2}\right)^2}} = BC$$

$$\frac{R}{\tau}\sqrt{4 - \frac{4}{6 + 2\sqrt{5}}} = BC$$

$$\frac{R}{\tau}\sqrt{4 - \frac{2}{3 + 2\sqrt{5}}} = BC$$

$$\frac{R}{\tau}\sqrt{4 - \frac{6 - 2\sqrt{5}}{4}} = BC$$

$$\frac{R}{\tau}\sqrt{\frac{10 + 2\sqrt{5}}{4}} = BC$$

$$\frac{R}{2\tau}\sqrt{10 + 2\sqrt{5}} = BC$$

$$\frac{R}{2}\sqrt{\frac{10 + 2\sqrt{5}}{\left(\frac{1 + \sqrt{5}}{2}\right)^2}} = BC$$

$$\frac{R}{2}\sqrt{\frac{40 - 8\sqrt{5}}{4}} = BC$$

$$\frac{R}{2}\sqrt{10 - 2\sqrt{5}} = BC$$

b) From part a and Section 7,

$$AB = \tau\left[\left(\frac{R}{2}\right)\sqrt{10 - 2\sqrt{5}}\right].$$

Therefore,

$$AB = \frac{R}{2}\left(\frac{1 + \sqrt{5}}{2}\right)\sqrt{10 - 2\sqrt{5}}$$

$$\frac{R}{2}\sqrt{\frac{(1+\sqrt{5})^2}{4}(10-2\sqrt{5})}$$

$$=\frac{R}{2}\sqrt{\frac{6+2\sqrt{5}}{4}(10-2\sqrt{5})}$$

$$=\frac{R}{2}\sqrt{\frac{40+8\sqrt{5}}{4}}$$

$$=\frac{R}{2}\sqrt{10+2\sqrt{5}}$$

c) We will show that the length of $\overline{AB}$ in the figure below is equal to $\frac{R}{\tau}$.

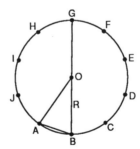

We are given that ABCDEFGHIJ is a regular decagon. Let O be the center of the circumscribed circle. Then $\overline{OB}$, is equal to R, which is the circumradius. If we draw $\overline{OA}$, then $\triangle OAB$ is a Golden Triangle. Hence, $\frac{R}{AB}=\tau$, or $\frac{R}{\tau}=AB=\frac{2R}{1+\sqrt{5}}$.

d) We will show that the length of $\overline{AD}$ in the figure below is equal to $R\tau$.

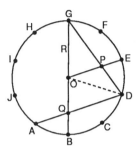

We are given that ABCDEFGHIJ is a regular decagon. Draw $\overline{GD}$ and $\overline{AD}$, two of the sides of the regular stardecagon. Also draw $\overline{GB}$, the diametral chord of the circumscribed circle. OG is equal to R, the length of the radial segment of the circumscribed circle. Draw $\overline{OE}$. The measure of central angle GOE is 72. The measure of angle GQD is 72. The measure of angle BGD is 36. Hence, $\triangle GQD \sim \triangle GOP$, and both triangles are Golden Triangles. Hence $\dfrac{GD}{QD} = \dfrac{GP}{PO} = \tau$, or $\dfrac{GD}{QD} = \dfrac{R}{PO} = \tau$. ($R$ = GP because $\triangle GOP$ is isosceles.) Hence, GD = QD $\cdot$ τ. But if we draw in $\overline{OD}$, we see that GO = OD = QD = R so that GD = $R \cdot \tau$. Now by Theorem 6, page 28, GD = AD; so AD = $R\tau$.

Project

The length of a longer side is τ; the rectangles are Golden Rectangles.

SECTION 8 (pages 61–74)

1   There will be 144 pairs of rabbits.

2   Eighth convergent:

$$\cfrac{1}{1+\cfrac{1}{1+\cfrac{1}{1+\cfrac{1}{1+\cfrac{1}{1+\cfrac{1}{1+\cfrac{1}{1+\cfrac{1}{1}}}}}}}} = \frac{21}{34} \approx .61765$$

Ninth convergent:

$$\cfrac{1}{1+\cfrac{1}{1+\cfrac{1}{1+\cfrac{1}{1+\cfrac{1}{1+\cfrac{1}{1+\cfrac{1}{1+\cfrac{1}{1+\cfrac{1}{1}}}}}}}}} = \frac{34}{55} \approx .61818$$

Tenth convergent:

$$\cfrac{1}{1 + \cfrac{1}{1 + \cfrac{1}{1 + \cfrac{1}{1 + \cfrac{1}{1 + \cfrac{1}{1 + \cfrac{1}{1 + \cfrac{1}{1 + \cfrac{1}{1}}}}}}}}} = \frac{55}{89} \approx .61798$$

3　By forming ratios of larger and larger consecutive Fibonacci numbers

4　Since $\tau = 1 + \dfrac{1}{\tau}$ (from Problem 5, Section 4), we can merely add 1 to the closer and closer approximations to $\dfrac{1}{\tau}$ that were obtained in Problem 3.

*5　When $n = 1$, $x = \dfrac{1}{\sqrt{5}}\left[\left(\dfrac{1 + \sqrt{5}}{2}\right)^{1} - \left(\dfrac{1 - \sqrt{5}}{2}\right)^{1}\right]$

$$= \dfrac{1}{\sqrt{5}}\left(\dfrac{2\sqrt{5}}{2}\right) = 1$$

When $n = 2$, $x = \dfrac{1}{\sqrt{5}}\left[\left(\dfrac{1+\sqrt{5}}{2}\right)^2 - \left(\dfrac{1-\sqrt{5}}{2}\right)^2\right]$

$= \dfrac{1}{\sqrt{5}}\left[\dfrac{1+2\sqrt{5}+5}{4} - \dfrac{1-2\sqrt{5}+5}{4}\right]$

$= \dfrac{1}{\sqrt{5}}\left(\dfrac{4\sqrt{5}}{4}\right) = 1$

When $n = 3$, $x = \dfrac{1}{\sqrt{5}}\left[\left(\dfrac{1+\sqrt{5}}{2}\right)^3 - \left(\dfrac{1-\sqrt{5}}{2}\right)^3\right]$

$= \dfrac{1}{\sqrt{5}}\left[\dfrac{1+3\sqrt{5}+15+5\sqrt{5}}{8} - \dfrac{1-3\sqrt{5}+15-5\sqrt{5}}{8}\right]$

$= \dfrac{1}{\sqrt{5}}\left(\dfrac{16\sqrt{5}}{8}\right) = 2$

When $n = 4$, $x = \dfrac{1}{\sqrt{5}}\left[\left(\dfrac{1+\sqrt{5}}{2}\right)^4 - \left(\dfrac{1-\sqrt{5}}{2}\right)^4\right]$

$= \dfrac{1}{\sqrt{5}}\left[\dfrac{1+4\sqrt{5}+30+20\sqrt{5}+25}{16}\right.$

$\left. - \dfrac{1-4\sqrt{5}+30-20\sqrt{5}+25}{16}\right]$

$= \dfrac{1}{\sqrt{5}}\left(\dfrac{48\sqrt{5}}{16}\right) = 3$

We have generated the first four terms of the Fibonacci sequence; 1, 1, 2, 3.

6   **a)** $z = \dfrac{3 + \sqrt{13}}{2} \approx 3.302775$

    b) $3,\ 3 + \dfrac{1}{3},\ 3 + \dfrac{1}{3 + \dfrac{1}{3}},\ 3 + \dfrac{1}{3 + \dfrac{1}{3 + \dfrac{1}{3}}},$

$$3 + \cfrac{1}{3 + \cfrac{1}{3 + \cfrac{1}{3}}} \quad, \qquad 3 + \cfrac{1}{3 + \cfrac{1}{3 + \cfrac{1}{3 + \cfrac{1}{3}}}} \quad,$$

$$3 + \cfrac{1}{3 + \cfrac{1}{3 + \cfrac{1}{3 + \cfrac{1}{3 + \cfrac{1}{3}}}}}$$

The decimal equivalents are 3.000000, 3.333333, 3.300000, 3.303030, 3.302752, 3.302777, 3.302775

7   $ax^2 + bx + c = 0$

    $x^2 + \dfrac{b}{a}x + \dfrac{c}{a} = 0$

Let $r = \dfrac{b}{a}$ and $s = \dfrac{c}{a}$. Then $x(x + r) + s = 0$ or $x = \dfrac{-s}{r + x}$

$$x = \cfrac{-s}{r + \cfrac{-s}{r + \cfrac{-s}{r + \cdots}}}$$

*8   a) $\sqrt{1} = 1,\ \sqrt{1 - \sqrt{1}} = 0,\ \sqrt{1 - \sqrt{1 - \sqrt{1}}} = 1,$

$$\sqrt{1 - \sqrt{1 - \sqrt{1 - \sqrt{1}}}} = 0,$$

$$\sqrt{1 - \sqrt{1 - \sqrt{1 - \sqrt{1 - \sqrt{1}}}}} = 1$$

$$\sqrt{1 - \sqrt{1 - \sqrt{1 - \sqrt{1 - \sqrt{1 - \sqrt{1}}}}}} = 0$$

The values of the convergents seem to oscillate between 0 and 1, and thus do not approach a limiting value.

b) $x^2 - x - 1 = 0$

$x^2 = x + 1$

$x = \sqrt{x + 1}$

$x = \sqrt{1 + x}$

Therefore, $x = \sqrt{1 + \sqrt{1 + \sqrt{1 + \sqrt{1 + \sqrt{1 + \ldots}}}}}$

$\sqrt{1} = 1, \sqrt{1 + \sqrt{1}} = \sqrt{2} \approx 1.41,$

$\sqrt{1 + \sqrt{1 + \sqrt{1}}} = \sqrt{1 + \sqrt{2}} = \sqrt{2.414214} \approx 1.55,$

$\sqrt{1 + \sqrt{1 + \sqrt{1 + \sqrt{1}}}} = \sqrt{1 + \sqrt{1 + \sqrt{2}}} =$

$\sqrt{1 + 1.554230} = \sqrt{2.554230} \approx 1.60$

This method of solution seems to work for $x^2 - x - 1 = 0$. The successive convergents *appear* to be approaching the value $\tau \approx 1.618$.

10 $(a + b)^0 = 1$

$(a + b)^1 = a + b$

$(a + b)^2 = a^2 + 2ab + b^2$

$(a + b)^3 = a^3 + 3a^2b + 3ab^2 + b^3$

$(a + b)^4 = a^4 + 4a^3b + 6a^2b^2 + 4ab^3 + b^4$

$(a + b)^5 = a^5 + 5a^4b + 10a^3b^2 + 10a^2b^3 + 5ab^4 + b^5$

11 **a)** The seventh through thirteenth rows of Pascal's triangle are:

```
1  6 15  20  15   6   1
1  7 21  35  35  21   7   1
1  8 28  56  70  56  28   8   1
1  9 36  84 126 126  84  36   9   1
1 10 45 120 210 252 210 120  45  10   1
1 11 55 165 330 462 462 330 165  55  11   1
1 12 66 220 495 792 924 792 495 220  66  12   1
```

$(a + b)^{12} = a^{12} + 12a^{11}b + 66a^{10}b^2 + 220a^9b^3 +$
$495a^8b^4 + 792a^7b^5 + 924a^6b^6 + 792a^5b^7 + 495a^4b^8 +$
$220a^3b^9 + 66a^2b^{10} + 12ab^{11} + b^{12}$

b) The sums of the elements along the diagonals of Pascal's triangle are Fibonacci numbers.

**12 By a theorem in geometry we know that in every triangle the sum of the lengths of any two sides is greater than the length of the third side. To prove that three different Fibonacci numbers cannot be the lengths of the sides of a triangle, we will consider two possibilities—either the Fibonacci numbers are consecutive or they are not consecutive.

I) By definition, a Fibonacci number is equal to the sum of the preceding two numbers. Therefore, three consecutive Fibonacci numbers cannot be the lengths of the sides of a triangle.

II) Consider three nonconsecutive Fibonacci numbers F_m, F_n, and F_p such that $F_m < F_n < F_p$. The following inequalities express the fact that F_m, F_n, and F_p are nonconsecutive: $F_m \le F_{n-1}$ and $F_{n+1} \le F_p$. Since F_{n-1} and F_n are consecutive Fibonacci numbers, $F_{n-1} + F_n = F_{n+1}$. But $F_m \le F_{n-1}$, so upon substituting we see that $F_m + F_n \le F_{n+1}$. Also, since $F_{n+1} \le F_p$, it follows that $F_m + F_n \le F_p$, which cannot occur if F_m, F_n, and F_p are the lengths of the sides of a triangle. Hence, three nonconsecutive Fibonacci numbers cannot be the lengths of the sides of a triangle. Thus, in either case, we see that it is impossible for three Fibonacci numbers to be the side lengths of a triangle.

**13 Since $\dfrac{a}{b} < \dfrac{c}{d}$, where b and d are not 0, then $ad < bc$. Therefore, $ad + ab < cb + ab$, or $a(b + d) < b(a + c)$, or $\dfrac{a}{b} < \dfrac{a + c}{b + d}$. Similarly, $\dfrac{a}{b} < \dfrac{c}{d}$, where $b, d \ne 0$, implies that $ad < bc$. Hence, $ad + cd < bc + cd$, or $(a + c)d < (b + d)c$, or $\dfrac{a + c}{b + d} < \dfrac{c}{d}$. Since $\dfrac{a}{b} < \dfrac{a + c}{b + d}$ and $\dfrac{a + c}{b + d} < \dfrac{c}{d}$, it follows that $\dfrac{a}{b} < \dfrac{a + c}{b + d} < \dfrac{c}{d}$.

**14 This problem can be thought of as a special case of Problem 13. This is because any two consecutive terms are of the form $\dfrac{a}{b}$ and $\dfrac{c}{d}$ and are unequal. Suppose that $\dfrac{a}{b} < \dfrac{c}{d}$. Examination of the sequence reveals that the numerator of each term of the sequence is determined by adding the numerators of the two immediately preceding terms. Similarly, the denominator of each term is found by adding the denominators of the two immediately preceding terms. By Problem 13, however, such a term must be between the two immediately preceding terms.

SECTION 9 (pages 75–82)

1 *Example 1:* $F_2 = 1$, $F_3 = 2$, $F_4 = 3$
    $$2^2 - 1 \cdot 3 = 4 - 3 = 1$$
   Example 2: $F_6 = 8$, $F_7 = 13$, $F_8 = 21$
    $$13^2 - 8 \cdot 21 = 169 - 168 = 1$$
   Example 3: $F_{12} = 144$, $F_{13} = 233$, $F_{14} = 377$
    $$233^2 - 144 \cdot 377 = 54{,}289 - 54{,}288 = 1$$

2 If points A, B, C, and D were collinear, then $m \angle ABD$
   would be 180. Therefore, to prove that these points are
   not collinear it is sufficient to show that $m \angle ABD \neq$
   180. We will use the figures below in the proof. Notice
   that $m \angle ABD = m \angle 1 + m \angle 2$.

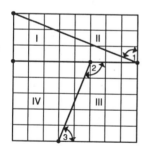
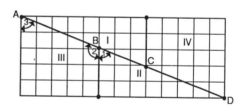

The following steps show that $m \angle 1 + m \angle 2 \neq 180$.

Arctan $\angle 3 = \arctan \left(\dfrac{5}{2}\right) \approx 68.2$.

Arctan $\angle 1 = \left(\dfrac{8}{3}\right) \approx 69.45$.

Since $m \angle 2 + m \angle 3 = 180$ and $m \angle 3 = 68.2$, it follows
that $m \angle 2 = 180 - 68.2 = 111.8$. Hence, $m \angle 1 + m \angle 2 =$
$69.45 + 111.8 = 180.53 \neq 180$.

3 According to the theorem in this section, we know that the area of parallelogram ABDC in Figure 16 is 1. Since the area of a parallelogram is equal to the product of the base and the altitude, we have $1 = a \cdot$ BD, in which BD $= \sqrt{F_{2n}^2 + (F_{2n-2})^2}$. So $a = 1 \sqrt{F_{2n}^2 + (F_{2n-2})^2}$. It is clear that as F_{2n} and F_{2n-2} get larger and larger, a gets smaller and smaller, and the parallelogram ABDC becomes less and less noticeable.

4 4, 6, 10: $4 \cdot 10 - 6^2$ $= 40 - 36$ $= 4$
   6, 10, 16: $6 \cdot 16 - 10^2$ $= 96 - 100$ $= -4$
   10, 16, 26: $10 \cdot 26 - 16^2$ $= 260 - 256$ $= 4$
   16, 26, 42: $16 \cdot 42 - 26^2$ $= 672 - 676$ $= -4$
   26, 42, 68: $26 \cdot 68 - 42^2$ $= 1{,}768 - 1{,}764 = 4$

Yes, the absolute value of the difference is the same in each case, namely 4.

5 The difference between the area of a square whose side length is the middle number of the consecutive triple and the area of a rectangle whose dimensions are the first and third numbers of the triple is 4 square units.

SECTION 10 (pages 83–90)

1 **a)** See the rectangle in Problem 4 of Section 7 that has $\overline{AF}$ as a side. This is a root-three rectangle.
  b) This is merely a rectangle whose side lengths are in the ratio 1:2.

2   The area of the square constructed on the longer side of the root rectangle is an *integral* multiple of the area of the square constructed on the shorter side of the root rectangle.

*3   **a)** Since $\overline{AG}$ forms a diagonal of ABML, it divides rectangle ABML into two congruent triangles that have equal areas.

    b) Since $\overline{AG}$ forms a diagonal of CMPG, it divides rectangle CMPG into two congruent triangles that have equal areas.

    c) The areas must be equal.

    d) Since $\triangle AEG \cong \triangle GDA$ and since $\triangle ABM \cong \triangle MLA$ and $\triangle MPG \cong \triangle GCM$, it follows from appropriate additions and subtractions of the areas associated with the above triangles that the area of DCML equals the area of MPEB.

    e) Adding the equal areas of DCML and MPEB to the area of ABML, we have immediately that the area of ABCD equals the area of AEPL.

4   ABCD is a unit square, so AB = 1 and BC = 1. Since M is the midpoint of $\overline{AB}$, then $MB = \frac{1}{2}$. Triangle BCM is a right triangle, hence, by the Pythagorean theorem $CM = \frac{\sqrt{5}}{2}$. However, MC = MR = MQ. So QR = 2MR = $\sqrt{5}$.

5   The area of a square with side $\overline{QR}$ is $(\sqrt{5})^2 = 5$, and the area of a square with side $\overline{BC}$ is $1^2 = 1$.

**6 We know that $\dfrac{QB}{BC} = \dfrac{BC}{BR}$. But QA = BR, QB = QA + AB

= BR + AB and AB = BC. By substitution, we obtain

$\dfrac{BR + AB}{AB} = \dfrac{AB}{BR}$. So $BR^2 + (AB)(BR) = AB^2$, or $1 + \dfrac{AB}{BR} =$

$\left(\dfrac{AB}{BR}\right)^2$, or $\left(\dfrac{AB}{BR}\right)^2 - \dfrac{AB}{BR} - 1 = 0$. This equation, however,

is an equation whose solution we know to be τ. Hence,

$\dfrac{AB}{BR} = \tau$. It follows that B divides $\overline{AR}$ into the Golden

Section.

*7 In Figure 17, $\triangle ABM \sim \triangle AEG$ (A.A.) Hence, $\dfrac{BM}{AB} = \dfrac{EG}{EA}$,

or $BM = \dfrac{(AB)(EG)}{EA}$. Since AB = 1, EG = 1, and

$AE = \sqrt{2}$, then $BM = \dfrac{1 \cdot 1}{\sqrt{2}} = \dfrac{1}{\sqrt{2}} = \dfrac{\sqrt{2}}{2} \approx \dfrac{1.414214}{2} \approx$

$.70707 \approx .7071$.

APPENDIX D (pages 105–114)

1 35 triangles:

AFT	NIJ	NPA	TNE	ENJ	TNI	FPN
TJN	EIH	EAT	NEP	NFT	ATJ	JEA
NIE	PHG	PTN	EPA	TGA	PAF	IPT
EHP	AGF	ANE	PAT	AHP	EPG	HAN
PGA	TFJ	TEP	ATN	PIE	NEH	GTE

2

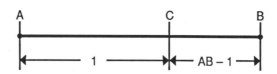

If AC = 1, then AB/AC = AC/CB becomes AB/1 = 1/AB − 1 or $AB^2 - AB - 1 = 0$. Using the quadratic formula, we find that AB = $(1 - \sqrt{5})/2$ or AB = $(1 + \sqrt{5})/2$. The negative value for AB, $1 - \sqrt{5}/2$, is disregarded. But AB = AB/1 = AB/AC (since AC = 1), so AB/AC = $1 + \sqrt{5}/2 = \tau$. But AB/1 = AB/AC (since AC = 1), so we see that AB/AC = $(1 + \sqrt{5})/2 = \tau$.

3 PROOF: In order to show that our construction procedure is correct, it will be enough to show that the central angle intercepted by $\overset{\frown}{DE}$ is 72°, that is, to show that $m \sphericalangle DBE = 72$.

In ΔBCD (by the Law of Cosines),

$$BD^2 = (BC^2 + CD^2 - 2(BC)(CD) \cos BCD$$

but

BC = AB/2

and

CD = BC,

so

CD = AB/2;

also

$$m \sphericalangle BCD = 180 - m \sphericalangle ACB$$

where

$$m \sphericalangle ACB = \tan^{-1} [AB/BC] = \tan^{-1} [AB/AB/2] = \tan^{-1} 2$$

or

$$m \sphericalangle ACB = 63.434949$$

So

$$m \sphericalangle BCD = 180 - 63.434949 = 116.56505$$

Substituting, we have

$$BD^2 = (AB/2)^2 + (AB/2)^2 - 2(AB/2)(AB/2) \cos 116.56505$$
$$BD^2 = AB^2/2 - (AB^2/2) \cos 116.56505$$
$$= (AB^2/2)(1 - \cos 116.56505)$$

So

$$BD = AB \sqrt{(1 - \cos 116.56505)/2}$$

Now in $\triangle BDE$ (again by the Law of Cosines):

$$DE^2 = BD^2 + BE^2 - 2(BD)(BE)(\cos DBE)$$

but

$$DE^2 = AB^2$$

and

$$BE = BD = AB \sqrt{(1 - \cos 116.56505)/2}.$$

Substituting in equation (1), we have:

$$AB^2 = 2(AB^2/2)(1 - \cos 116.56505) - 2(AB^2/2)$$
$$(1 - \cos 116.56505) \cos DBE$$
$$AB^2 = AB^2 (1 - \cos 116.56505) - AB^2$$
$$(1 - \cos 116.56505) \cos DBE$$

$$\frac{AB^2 - AB^2 (1 - \cos 116.56505)}{- AB^2 (1 - \cos 116.56505)} = \cos DBE$$

$$\frac{1 - (1 - \cos 116.56505)}{- (1 - \cos 116.56505)} = \cos DBE$$

$$\frac{\cos 116.56505}{- (1 - \cos 116.56505)} = \cos DBE$$

$$\frac{- 0.4472136}{- 1.4472136} = \cos DBE$$

$$0.309017 = \cos DBE$$

So

$$m \angle DBE = \cos^{-1} (0.309017)$$
$$= 72°$$

Q.E.D.

5   The numerical value of these ratios approaches τ.

7   Photocopy and cut out the pattern shown in Figure 37.
    Fold along the dotted lines. Use tape to secure the
    edges and form a golden cuboid. An examination of the
    solid should make it intuitively clear that all four
    internal diagonals (see AD, FC, BE, and GH in Figure
    35) are congruent and would be concurrent at a

A Golden Cuboid

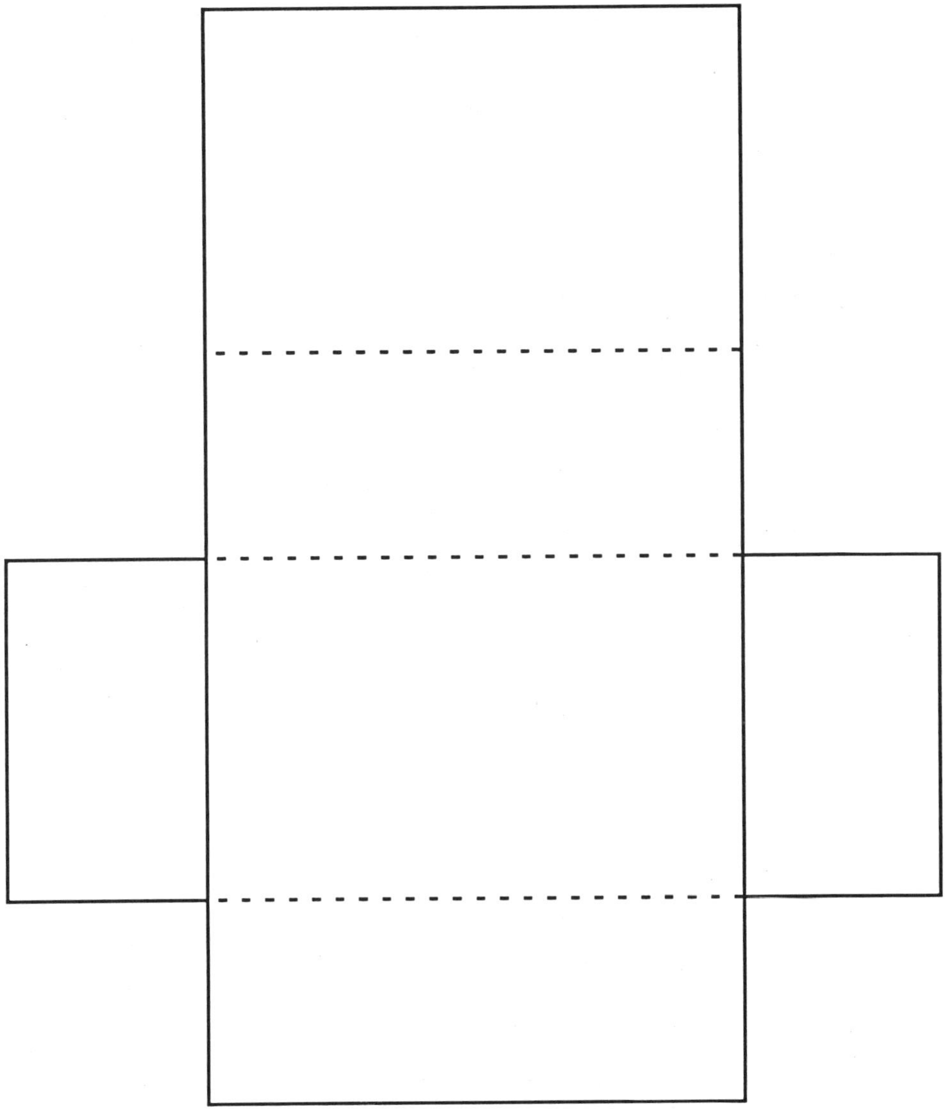

Figure 37

common midpoint. Seeing this makes it clear that a sphere could be imagined that would have the common midpoint of the diagonals as its center and that would contain all eight vertices on the cuboid.

A double application of the Pythagorean theorem allows us to calculate the length of each internal diagonal as 2. This follows since, for example,

$$AD = AB^2 + BC^2 + CD^2$$

or, upon substituting,

(1) $AD = \sqrt{1^2 + \tau^2 + (1/\tau)^2}$

Recall from Appendix D, Problem 2 the equation $AB^2 - AB - 1 = 0$, where AB was shown to have a value of τ. Knowing this means that $\tau^2 - \tau - 1 = 0$ or $\tau^2 = \tau + 1$. Furthermore, from Problem 5 in Section 4, we know that $\tau = (1/\tau) + 1$ or that $1/\tau = \tau - 1$. Making substitutions in equation (1) above, therefore, yields the following:

$$\begin{aligned} AD &= \sqrt{1^2 + (\tau + 1) + (\tau - 1)^2} \\ &= \sqrt{1 + \tau + 1 + \tau^2 - 2\tau + 1} \end{aligned}$$

Substituting again yields

$$\begin{aligned} &= \sqrt{2 + \tau + (\tau + 1) - 2\tau + 1} \\ &= 4 \\ &= 2 \end{aligned}$$

It follows that the radius of the sphere that circumscribes the golden cuboid is 1. Its surface area is then $4\pi(1)^2$ or 4π.

The surface area of the golden cuboid will be the sum of the areas of all six of its faces—that is, $2[\tau(1/\tau)]$ + 2τ + $2(1/\tau)$. Substituting and simplifying, we have the surface area being $2 + 2\tau + 2(\tau - 1)$ or $2 + 2\tau + 2\tau - 2$ or 4τ.

It follows at once, then, that the ratio of the surface area of the golden cuboid to that of its circumscribing sphere is $4\tau/4\pi$ or simply $\tau{:}\pi$, which is what we wished to prove.

8 Problem 5 in this appendix, along with our work in Section 8, suggests that as one calculates the ratio of adjacent terms in an additive sequence (i.e., a generalized Fibonacci sequence), one finds that as the size of the terms increases, the value of the ratio (formed by dividing any term by its predecessor) approaches the value of τ or $(1 + \sqrt{5})/2$. To find the answer to this problem, therefore, divide the last addition of water (1,000,000 gallons) by the preceding amount (call it z), and equate this ratio to $(1 + \sqrt{5})/2$. That is:

$$\frac{1,000,000}{z} = \frac{1 + \sqrt{5}}{2}$$

Solving for z (with the aid of a calculator) and rounding to the nearest integer will yield 618,034. Working backwards, one should be led to the first two amounts, and the first amount should be less than the second one. However, the solution to the problem is 154 gallons for the first amount and 144 gallons for the second amount—and the algorithm above does not produce it. However, one can use the values it produces for the nineteenth and twentieth additions of water to get a system of equations that does yield the solution.

Consider the sequence of amounts, if x represents the first number of gallons added and y represents the second one:

$$x, y, x + y, x + 2y, 2x + 3y, 3x + 5y, \ldots$$

The coefficients of x and y in these terms are Fibonacci numbers, and the sequence can be defined recursively by

$\text{deposit}_1 = x$
$\text{deposit}_2 = y$
$\text{deposit}_n = F_{n-2}\, x + F_{n-1}\, y$, where F_n represents the
   nth Fibonacci number.

The equations that describe the nineteenth and twentieth amounts of water are

$F_{17}\, x + F_{18}\, y = 618{,}034$
$F_{18}\, x + F_{19}\, y = 1{,}000{,}000$

Substituting the Fibonacci numbers gives

$1{,}597x + 2{,}584y = 618{,}034$
$2{,}584x + 4{,}181y = 1{,}000{,}000$

Solving this system yields $x = 154$ and $y = 144$. Find all twenty amounts of water added to the aquarium.

9   One way to solve this problem is to use the Exterior Angle Theorem, which says that the measure of any exterior angle of a triangle is equal to the sum of its two remote interior angles. It follows, therefore, that:

$$m \angle 7 = m \angle 1 + m \angle 9$$
$$m \angle 8 = m \angle 2 + m \angle 10$$
$$m \angle 11 = m \angle 3 + m \angle 12$$

$$m \angle 13 = m \angle 4 + m \angle 14$$
$$m \angle 15 = m \angle 5 + m \angle 6$$

$$m \angle 7 + m \angle 8 + m \angle 11 + m \angle 13 + m \angle 15 = (m \angle 1 + m \angle 2 + m \angle 3 + m \angle 4 + m \angle 5) + (m \angle 9 + m \angle 10 + m \angle 12 + m \angle 14 + m \angle 6)$$

But:

$$m \angle 7 = m \angle 8 + m \angle 11 + m \angle 13 + m \angle 15 = 540$$
(Why?)

and

$$m \angle 9 + m \angle 10 + m \angle 12 + m \angle 14 + m \angle 6 = 360$$
(Why?)

Substituting, then, we have:

$$540 = (m \angle 1 + m \angle 2 + m \angle 3 + m \angle 4 + m \angle 5) + 360$$

So,

$$180 = m \angle 1 + m \angle 2 + m \angle 3 + m \angle 4 + m \angle 5$$

Bibliography

ARTICLES AND BOOKS

Archibald, R. C. "The Golden Section." *American Mathematical Monthly* 25 (1918):232–238.

Ball, W. W. R. *Mathematical Recreations and Essays.* 11th ed. London, 1939:36–38.

Baravalle's H. V. "The Geometry of the Pentagon and the Golden Section." *The Mathematics Teacher* 41 (1948):22–31.

————. "The Golden Section and Fibonacci Numbers." *Scripta Mathematica* 16 (March–June, 1950):116–119.

Beatty, S. "Problem 3173." American Mathematical Monthly 33 (1926):159 and 34 (1927):159.

Becker, J. P. "On Solutions of Geometrical Constructions Utilizing the Compasses Alone." *The Mathematics Teacher* 57 (October 1964):398–403.

Bell, Eric Temple. "The Golden and Platinum Proportions." *National Mathematics Magazine* 19 (1944):20–26.

Bennett, A. A. "The Most Pleasing Rectangle." *American Mathematical Monthly* 30 (1923):27–30.

Bicknell, Marjorie and Verner E. Hoggatt, Jr. "The Golden Triangle." *The Fibonacci Quarterly* 7 (February 1969).

Boles, Martha and Rochelle Newman. *The Golden Relationship Art Math Nature Book 1: Universal Patterns.* Bradford, MA: Pythagorean Press, 1987.

Borissavelievitch, M. *The Golden Number and the Scientific Aesthetics of Architecture.* New York: Philosophical Library, 1958.

Bowes, Julian. "Dynamic Symmetry." *Scripta Mathematica* 1 (1933):236–244, 309-314.

Bragdon, Claude. "Observations on Dynamic Symmetry." In
Old Lamps for New. New York: Knopf, 1925.

Carnahan, Walter. "Compass Geometry." *School Science
and Mathematics* 32 (April 1932):384–390.

———. "Geometric Constructions Without the Compasses."
School Science and Mathematics 36) February
1936):182–189.

Courant, Richard and Herbert Robbins. *What is
Mathematics?* New York: Oxford University Press,
1941.

Coxeter, H. S. M. "The Golden Section, Wythoff's Game."
Scripta Mathematica 19 (1953):135–143.

———. Chapter 11, *Introduction to Geometry.* New York:
John Wiley & Sons, Inc., 1961.

Cundy, H. Martyn and A. P. Rollett. *Mathematical Models.*
revised ed. Oxford: Oxford University Press, 1951.

Dantzig, Tobias. *The Bequest of the Greeks.* New York:
Charles Scribner's Sons, 1955.

De Sales McNabb, Mary. "Phyllotaxis." *The Fibonacci
Quarterly* 1 (December 1963):57–60.

DeTemple, Duane. "Simple Constructions for the Regular
Pentagon and Heptadecagon." *Mathematics Teacher* 82
(May 1989):356–365.

Dudeney, Henry Ernest. *536 Puzzles and Curious
Problems.* New York: Charles Scribner's Sons, 1967.

Eves, Howard & Carroll V. Newsom. *An Introduction to the
Foundations and Fundamental Concepts of
Mathematics.* New York: Holt, Rinehart & Winston,
1958.

Gardner, Martin. "Math Games." *Scientific American*
(August 1959): 128–134.

———. *The Second Scientific American Book of
Mathematical Puzzles and Diversions.* New York: Simon
and Schuster, 1961.

————. *Wheels, Life and Other Mathematical Amusements.* New York: W. H. Freeman and Co., 1983.

Garland, Trudi H. *Fascinating Fibonaccis: Mystery and Magic in Numbers.* Palo Alto, CA: Dale Seymour Publications, 1987.

Ghyka, Matila. *The Geometry of Art and Life.* New York: Sheed and Ward, 1946.

Graesser, R. F. "The Golden Section." *The Pentagon* 3 (1943–44):7–19.

Grunbaum, Branko. "Geometry Strikes Again." *Mathematics Magazine* 58 (January 1985):12–18.

Gugle, Marie. "Dynamic Symmetry." *Third Yearbook.* National Council of Teachers of Mathematics (1928):57–64.

Hallerberg, Arthur. "The Geometry of the Fixed-Compass." *The Mathematics Teacher* 52 (April 1959):230–244.

Hambidge, Jay. *Practical Applications of Dynamic Symmetry.* New York: The Devin-Adair Company, 1932.

————. *The Elements of Dynamic Symmetry.* New Haven: Yale University Press, 1923.

————. *Dynamic Symmetry in Composition.* New Haven: Yale University Press, 1923.

Heath, Sir Thomas L. *The Thirteen Books of Euclid's Elements.* 3 vols. New York: Dover, 1956.

Hess, A. L. "Certain Topics Related to Constructions with Straight Edge and Compasses." *Mathematics Magazine* 29 (March–April 1956):217–221.

Hlavaty, Julius H. "Macheroni Constructions." *The Mathematics Teacher* 50 (November 1957):482–487.

Hoggatt, Verner E., Jr. *Fibonacci and Lucas Numbers.* Boston: Houghton Mifflin Company, 1969.

Holt, Marvin. "Mystery Puzzler and Phi." *The Fibonacci Quarterly* 3 (April 1965):135–138.

————. "The Golden Section." *The Pentagon* (Spring 1964): 80-104.

Huntley, H. "The Golden Cuboid." *The Fibonacci Quarterly* 2 (October 1964):184.

Huntley, H. *The Divine Proportion: A Study in Mathematical Beauty*. Mineola, NY: Dover Publications, 1970.

Karchmar, E. J. "Phyllotaxis." *The Fibonacci Quarterly* 3 (February 1965):64–66.

Lamb, John F. Jr. "The Rug-Cutting Puzzle." *Mathematics Teacher* 80 (January 1987):12–13.

Law, Bernadine. "Some Mathematical Considerations of Dynamic Symmetry Underlying Visual Art." *The Pentagon* 13 (1953)7–17.

Olds, C.D. *Continued Fractions*. New York: Random House, 1963.

Olson, Alton T. *Mathematics Through Paper Folding*. Reston, VA: 1975.

Pennick, Nigel. *Sacred Geometry*. San Francisco: Harper and Row, 1980.

Raab, Joesph. "The Golden Rectangle and Fibonacci Sequence, as Related to the Pascal Triangle." *The Mathematics Teacher* 55 (November 1962):538–543.

Reagan, James. "Thar's Gold in Them Thar Conic Sections." *Mathematics Teacher* 77 (May 1984):357–359.

Sarton, George. *Introduction to the History of Science*. (3 vols.) Baltimore: The Williams and Wilkins Co., 1931.

Satterly, John. "Meet Mr. Tau." *School Science and Mathematics* 56 (1956):731–741.

————. "Meet Mr. Tau Again." *School Science and Mathematics* 57 (1957):150–151.

Schielack, Vincent P. Jr."The Fibonacci Sequence and the Golden Ratio." *Mathematics Teacher* 80 (May 1987):357–358.

Seitz, Donald T., Sr. "A Geometric Figure Relating the Golden Ratio and Pi." *Mathematics Teacher* 79 (May 1986):340–341.

Mathematics Teacher, The. 43 (1950). This reference contains a large number of books and articles related to mathematics and art.

Tietze, Heinrich. *Famous Problems of Mathematics*. New York: Graylock Press, 1965.

Thompson, Sir D'Arcy W. *On Growth and Form.* Cambridge: The University Press, 1942.

Vorob'ev, N. N. *Fibonacci Numbers.* New York: Blaisdell Publishing Company, 1961.

Zippin, Leo. *Uses of Infinity*. New York: Random House, 1962.

FILMS

DONALD IN MATHMAGIC LAND. This is a 1959 Walt Disney Production. This 26-minute color film is appropriate for almost any age level and does not require a strong math background. It is available in video cassette format and may be obtained from many video outlets.

THE GOLDEN SECTION. This 30-minute film is geared at a significantly higher level than is *Donald in Mathmagic Land.* It is available from Audio Brandon Films, Inc., 34 Mac Questen Parkway South, Mount Vernon, New York 10550, (914) 664-5051.

Index